James E. Christie

From Influence to Inhabitation

The Transformation of Astrobiology in the Early Modern Period

James E. Christie
Sydney, Australia

ISSN 0066-6610 ISSN 2215-0307 (electronic)
International Archives of the History of Ideas Archives internationales d'histoire des idées
ISBN 978-3-030-22171-3 ISBN 978-3-030-22169-0 (eBook)
https://doi.org/10.1007/978-3-030-22169-0

© Springer Nature Switzerland AG 2019
This work is subject to copyright. All rights are reserved by the Publisher, whether the whole or part of the material is concerned, specifically the rights of translation, reprinting, reuse of illustrations, recitation, broadcasting, reproduction on microfilms or in any other physical way, and transmission or information storage and retrieval, electronic adaptation, computer software, or by similar or dissimilar methodology now known or hereafter developed.
The use of general descriptive names, registered names, trademarks, service marks, etc. in this publication does not imply, even in the absence of a specific statement, that such names are exempt from the relevant protective laws and regulations and therefore free for general use.
The publisher, the authors, and the editors are safe to assume that the advice and information in this book are believed to be true and accurate at the date of publication. Neither the publisher nor the authors or the editors give a warranty, express or implied, with respect to the material contained herein or for any errors or omissions that may have been made. The publisher remains neutral with regard to jurisdictional claims in published maps and institutional affiliations.

This Springer imprint is published by the registered company Springer Nature Switzerland AG.
The registered company address is: Gewerbestrasse 11, 6330 Cham, Switzerland

To my parents, John and Kathryn

Preface

It is often said that 'Are we alone in the universe?' is one of humanity's oldest and most profound questions. The problem with this statement, as a student of intellectual history might point out, is that the concepts of 'we', 'alone' and 'universe' are not culturally or temporally constant. We know what *we* mean by the question. Is there life (as we know it) somewhere other than on the earth? This question has a relatively long history, and it's a history that has been gaining more attention since new discoveries—the detection of planets orbiting around other stars and of living creatures surviving in extreme environments here on earth—reignited scientific and popular interest in the possibility of alien life. Journalistic platforms are now wont to publicise any scientific findings that give even a hint of potential extraterrestrial habitability. On some of these platforms, you can also read your horoscope.

The idea for this book arose when, with a background in the history of astrology, I began to read about the history of the extraterrestrial life debate. As I studied this history, I began to perceive connections, both abstract and specific, productive and conflicting, between the traditions of astrology and cosmic pluralism, and thus a research project was born. This book is the first result of that research. It presents a long view of the historical interactions between the astrological tradition and the 'plurality of worlds' philosophy from the classical period up to the Enlightenment, with a focus on the period 1580–1700. It is not intended to be, nor could it be, exhaustive in terms of either the instances or the forms of these interactions. It is intended rather to provide a historical and methodological framework, demonstrating the advantages of studying these traditions in tandem and making suggestions towards a historical narrative that future research may develop or indeed disrupt.

To that end, the two main theses of the work attempt to establish that there were causal influences linking the opposite trajectories of astrology and pluralism in the long seventeenth century. These connections have been seldom studied or appreciated by historians of either discipline, yet they are pertinent to both. Readers interested in the history of astrology, for example, will hopefully come to appreciate the relevance of the 'plurality of worlds' philosophy to their discipline, e.g. how ideas about extraterrestrial life could be both augmentative and restrictive to models of celestial influence. Many other questions are raised along the way, concerning such subjects as celestial motions and activities, the uniformity and diversity of nature, the cosmic status of humanity and the varying associations of both astrology and pluralism to the issue of atheism. On all these fronts, I hope this book will spark debate and invite further research and analysis.

Today, scientific engagement with the question of extraterrestrial life falls within the discipline of astrobiology. As a discipline, its history is short, having been founded only in the 1990s. In terms of its subject matter, however, the history of astrobiology is often considered synonymous with that of the extraterrestrial life debate. Thus, its point of origin is usually traced to the astronomical revolution of the sixteenth and seventeenth centuries. Readers more concerned with this history will discover a new side to the story, i.e. the various ways that this early development was affected by the astrological tradition within which and against which it evolved.

In the view of some, the current astrobiological revolution encompasses more than just the science. It represents a striving for a more complete cosmology, a true 'cosmic perspective' that can answer the question of how terrestrial life and humanity fit into the universal picture. This is why I decided to subtitle the book 'The Transformation of Astrobiology in the Early Modern Period'. There was, of course, no such thing as early modern astrobiology. The discussion on this front is thus largely confined to sections in the introduction and conclusion where historicity is briefly subordinated to a historian's attempt at anthropology. Those who are interested will hopefully find food for thought concerning historical and contemporary confluences of astronomy and biology. Others who find such anachronisms objectionable may wish to skip over those sections and enjoy what is, essentially, a history of early modern natural philosophy.

Many individual authors and philosophers are discussed in this book, and in most cases, novel contributions are made to the study of these figures per se. In some instances, though, it is simply a case of looking at these sources through the compound lens of astrology and pluralism. Despite the broad chronology, an attempt is still made where possible to provide a thick history, with long quotations—especially when there is no available English translation—and original language in the footnotes. This is done with the intention that the book might serve as a useful resource to a wide range of readers.

There are many people requiring recognition and thanks for the role they played in the making of this work. First of all, I must acknowledge all the scholars whose work laid the foundation for my own. I am especially indebted to Guido Giglioni and Charles Burnett for their expert guidance and encouragement. I am also grateful for the advice received from colleagues, too numerous to mention, at The Warburg Institute and elsewhere. And thanks must go, of course, to Jasmin and to my family and friends for their unwavering support and belief.

Sydney, Australia James E. Christie

Contents

Chapter 1
Introduction: Astrology, Extraterrestrial Life and Astrobiology

Astrobiology recognises that it is difficult to develop a full understanding of life on Earth without understanding its links to the cosmic environment.

(Cockell 2015, 4)

Abstract The history of astrology and the history of the extraterrestrial life debate are largely kept separate, for reasons both chronological and historiographical. In fact, there was a large period of overlap in which many historical actors seriously considered and wrote on the concepts of both celestial influence and celestial inhabitation. The history of astrobiology, understood broadly as the study of 'life in a cosmic context', offers a potential avenue for exploring the relationships between astrology and cosmic pluralism in the early modern period. The two main theses of the book are presented: (a) that evolving theories of celestial influence in the sixteenth and seventeenth centuries encouraged speculation about extraterrestrial life, and (b) that as the seventeenth century progressed, certain thinkers began to consciously oppose the concepts of influence and inhabitation as rival teleological models for astronomical cosmology, with the latter emerging triumphant. The rest of the chapter comprises a literature review, a section on terminology, and a section on methodology, assessing the usefulness of Kuhn's 'paradigms' and Lovejoy's 'unit-ideas' in approaching the subject matter.

Keywords Astrology · Astrobiology · Extraterrestrial life · Plurality of worlds · Pluralism · Teleology

© Springer Nature Switzerland AG 2019

J. E. Christie, *From Influence to Inhabitation*, International Archives of the History of Ideas Archives internationales d'histoire des idées 228,
https://doi.org/10.1007/978-3-030-22169-0_1

1.1 The Aim of the Work

This book presents a conjoined and comparative history of astrology and the philosophical tradition related to the idea of a 'plurality of worlds' in the early modern period. In doing so, it suggests new research areas for both histories, as well as a potential new direction for the history of astrobiology, the modern scientific discipline which encompasses the search for extraterrestrial (ET) life. The idea that life, in a form some way comparable to life on earth, may exist on another celestial body has been a subject of great scientific and cultural interest for the past several centuries. Prior to the seventeenth century, however, this idea was a minority view within the Western intellectual tradition. In medieval and early modern cosmology, the celestial bodies were more often understood in terms of the influences they were thought to exert upon the terrestrial realm. The sixteenth and seventeenth centuries saw the decline of the monopoly of Aristotelian physics and Ptolemaic astronomy in the domain of Western natural philosophy. The astronomical discoveries and innovations of such figures as Galileo, Kepler and Newton laid the foundations for what would become our modern understanding of the universe and the place of the earth within it. Part of this transition involved the decline or marginalisation of astrology, the dismantling of the celestial causal chain of Aristotelian cosmology and the dismissal of any planetary or astral influence outside of light, heat and gravitation.[1] Another part involved the adoption, or re-adoption, of the cosmological view concerning the existence of a plurality of worlds in the universe (hereafter 'pluralism'). In this instance, pluralism refers to the belief that some or all of the celestial bodies are similar in nature to the earth and inhabited by living creatures.[2]

These two phenomena are usually dealt with separately in historical scholarship, largely because the same period represents the *terminus ad quem* for the serious study of astrology and the *terminus a quo* for the growing acceptance of pluralism. Nevertheless, both trends are considered as consequences of a Copernican cosmology and hallmarks of a modern worldview. The modern delineation between these two strands of historical enquiry (i.e. the history of astrology and the history of pluralism) may be detrimental, both to our understanding of celestial philosophy at any given time, as well as to our appreciation of cosmological change over longer periods. While it is of course a truism to say that early modern science or natural philosophy was more holistic than disciplinary, there are particularly good reasons to consider astrology and pluralism in tandem. The most general and obvious similarity is that both concepts meld astronomy and the life sciences. Astrology is concerned with the effect of the celestial realm on terrestrial phenomena, including biological generation (celestial influence), while pluralism is predicated on the possible existence of biological processes in the heavens (celestial inhabitation).

---

[1] For recent engagements with this topic, see the various contributions in Granada et al. 2016, and also Vermij and Hirai 2017. See also Whitfield 2001, 177–87; Rutkin 2006, 2018; Vermij 2014.

[2] The main monograph works on this subject are Lovejoy 1948; Dick 1982; Guthke 1983; Crowe 1986.

Aristotle, in his work *On the Parts of Animals*, twice compared the science of the stars to the science of living objects. The first comparison argues that mortal animals, just as much as the heavens, are products of a defining cause or principle (*PA*, I.1, 641b13–21). Actions in the celestial and the biological world can be investigated and understood in the same way, according to the theory that 'whenever there is plainly some final end, to which a motion tends should nothing stand in the way, we always say that such final end is the aim or purpose of the motion' (*PA*, I.1, 641b24–26; Aristotle 1882, 8). That is, both sciences can be approached teleologically. Aristotle then juxtaposed the two sciences: the heavens are more divine than plants or animals, yet our knowledge of the latter is far more certain. The two sciences thus complement each other. The 'greater nearness and affinity to us' of living creatures balance 'the loftier interest of the heavenly things that are the objects of the higher philosophy' (*PA*, I.5, 645a1–4; Aristotle 1882, 16).[3] Aristotle even acknowledges the less glamourous nature of biology when compared to astronomy, but reassures the philosophical reader that it will provide similar pleasures by revealing the links of causation and the artistic spirit that underlies all things.

In the mid 1990s, a new discipline was founded which similarly unites astronomy and biology under a common goal, in which the certainty and nearness-to-hand of the one balance the (arguably) more speculative nature of the other. This discipline is astrobiology, a flourishing science which has largely replaced the field of 'exobiology', an earlier disciplinary label coined by Joshua Lederberg (1925–2008) in 1960 to describe the scientific search for ET life.[4] Astrobiology operates through a variety of research projects, funded primarily by NASA, and as a scientific discipline it benefits from broad appeal and exposure, especially in regards to the search for potentially habitable exoplanets and, indeed, the not yet exhausted search for life within our solar system.[5] With the development of a science comes, inexorably, the history of that science. This decade has seen a concerted attempt to establish the history and philosophy of astrobiology 'as a research field in its own right' (Dunér et al. 2013, 3). This history, when it searches for origins prior to the space age, has largely been equated with the history of the 'plurality of worlds' philosophy and the ET life debate. If we consider historical usages of the word 'astrobiology', however, as well as the modern disciplinary description, we can find considerable scope to expand the history of astrobiology, and pluralism in general, into related fields such as astrology.[6]

According to the Oxford English Dictionary, astrobiology was a term first used in publications put out by the Koreshan Unity, an American sect which followed the

---

[3] On this topic, see also Leunissen 2009.

[4] For the modern history of astrobiology, see Dick and Strick 2005.

[5] NASA operates in tandem with many partner agencies and organisations. As a starting point, see the NASA Astrobiology website. Homepage, Astrobiology, https://astrobiology.nasa.gov/ [accessed 10 November 2017].

[6] There have been some recent attempts to look at the history of other aspects of astrobiology, such as theories of panspermia and the origins of life more generally. However, they contain little discussion of the early modern period. See Temple 2007; Demets 2012; Dunér et al. 2016.

teachings of the eclectic nineteenth-century physician and alchemist, Cyrus Teed (1839–1908).[7] In his usage, astro-biology was a science which studied the 'regulation of human affairs by the clock-work of the cosmos' (Teed 1908). Around the same time, the French philosopher Henry Lagrésille, active at the turn of the century, used the term in his metaphysical work *Le fonctionnisme universel* ('On Universal Functionism', Lagrésille 1902). Lagrésille, who in this work described functions from the atomic to the stellar level as creations of the free activity of spirit, used the word astrobiology to mean a 'qualitative law of energy' (*loi qualitative de l'énergie*) (Lagrésille 1902, 540–41). 'I am afraid that neither the philosopher nor the scientist will care much for this book', remarked one reviewer, 'but the theosophist may find it edifying' (Morrison 1903).

In the 1930s the term was used with a different meaning by the French historian René Berthelot (1872–1960), who used it to describe an intermediate stage in human cosmological development. In between 'savage animism and modern science', astrobiology combined a vitalistic (or anthropomorphic) interpretation of the heavens with a desire to formulate laws that governed their motion and influence on terrestrial phenomena (Philosophical Periodicals 1934, 269; Berthelot 1938). All these historical usages are more closely connected to astrology than to the search for ET life. It was only in the mid-twentieth century that it took on this new meaning. It was used in several published works and astronautical conferences in a sense synonymous to Lederberg's 'exobiology', or a similar now-disused term 'xenobiology' (Catling 2013, 4). Yet in French it retained the meaning given by Berthelot, and it was used by the historian of philosophy Paul-Henri Michel (1894–1964) to describe Giordano Bruno's animistic cosmology (Michel 1973, 216). Then, in the 1990s, it took on its current scientific meaning, defined by NASA as the study of the origins, evolution, distribution, and future of life in the universe.[8] Astrobiology now denotes an expanded and interdisciplinary science, combining the search for ET life with the study of terrestrial biology—especially its origins, evolution, and occurrence in extreme environments—in an attempt to understand the very nature of life itself within a larger cosmological context.

It is in this shorter and more general definition, 'the study of life in a cosmic context' (Catling 2013, 2), that astrobiology is most obviously connected to the astrological tradition. Consider this passage from a textbook on astrobiology:

> Astrobiology recognises that it is difficult to develop a full understanding of life on Earth without understanding its links to the cosmic environment. The Earth seems like a tranquil place. However, it is subjected to the vagaries of its astronomical environment. For example, a leading hypothesis for the extinction of the dinosaurs is an asteroid or comet impact about 65 million years ago. This hypothesis underscores the fact that to understand past life on Earth we need to understand how the astronomical environment may have influenced life (Cockell 2015, 4).

Leaving aside mention of dinosaurs, this quote could easily be used as an astrological manifesto from the ancient, medieval or early modern periods. Indeed, the

[7] 'astro-, comb. form', *OED Online*. Oxford University Press [accessed 12 January 2017].

[8] The French still use the term 'exobiologie'.

case of astrology illustrates a simple yet profound point: the modern science of astrobiology unites two fields of enquiry, astronomy and biology, which in fact only became separated following the Scientific Revolution in the early modern period, or perhaps even later still, following the disciplinisation efforts of William Whewell (1794–1866).[9] The importance of astrology to pre-modern medicine is the most palpable example of this unification, but the effect of the heavens on biological processes had much wider scientific, philosophical, agricultural and cultural ramifications.

All this is to say, that the history of astrobiology may provide the perfect vehicle to connect the more established histories of astrology and pluralism. The historical coincidence of meanings, with astrobiology being used in different senses synonymous with astrology, presents an opportunity to write a deeper contextual history of attempts to understand life in reference to perceptions of the cosmos.[10] This is of course borrowed territory from what we might call the history of 'cosmology' or 'cultural astronomy'. Framing it as a history of astrobiology, a modern field, is an attempt to create a compromise between relevance and relativism—to risk anachronism in the hope of drawing out meaningful associations between modern and pre-modern pursuits of knowledge.[11] It is a history of astrobiology purely in an abstracted, interpretative sense, intended both to contextualise the historical narrative to which modern astrobiology lays claim, and to suggest a form of continuity underlying the astronomical upheavals of the early modern period. The more concrete, historical aim of this research, meanwhile, is to probe the links between theories of celestial influence and celestial inhabitation in early modern natural philosophy. There is rich ground for enquiry, due to the simple fact that most figures in the early history of pluralism were either practicing astrologers or wrote about the nature of celestial influence. These natural philosophers first began to extend generation and corruption to the heavens at a time when those very processes were still causally tied to celestial influence.

The results of research conducted thus far have led to two main theses. The first argument is that, in its infancy, pluralism was in fact encouraged by evolving astrological theories. This evolution was stimulated by elements of Platonic and Stoic philosophy as well as by post-Copernican reforms. A belief in ET life was thus nurtured by concepts of the ubiquity and necessity of celestial influence which survived, at least for a time, the dismantling of the Aristotelian/Ptolemaic cosmos. The twin notions of influence and inhabitation could combine to provide significant explanatory power for the cosmological alternatives to Aristotle that arose in the time

[9] See Sandoz 2016. Astrobiology prides itself on its trans-, multi- and/or interdisciplinary nature. See Santos et al. 2016.

[10] This is not to say that the use of 'astrobiology' as a disciplinary title implies that it has anything in common *per se* with historical or contemporary usages that are largely incongruent. The point is rather that the combination of the two roots 'astro' and 'bio' can signify various different ideas and activities, and we might draw interesting comparisons between them.

[11] On this point, see Chalmers 2016, 28: 'Our current knowledge provides us with a way of putting questions to history, a strategy that need not be problematic provided there are ways of ensuring that it is history that provides the answers.'

between Nicholas of Cusa and Newton, especially in terms of the motion and order of the celestial bodies. This argument will be the focus of Chaps. 2, 3, and 4. Chapter 2 takes the form of a broad survey of the interplay of astrological and pluralist ideas from Plutarch to Giordano Bruno, while Chaps. 3 and 4 are case studies of William Gilbert (1544–1603) and Johannes Kepler (1571–1640) respectively. The point to be made is that celestial influence and celestial inhabitation were not mutually exclusive concepts. In the attempt to build a new cosmology as robust and descriptive as that of Aristotle and Ptolemy, philosophers in the early modern period used both ideas, separately or together, to define a teleologically satisfying world-view.

This symbiosis of astrological and pluralist themes will be further explored in Chap. 5, which again takes the form of a broad survey, this time for the period between Kepler and Newton. This is partly a non-argumentative exploration of the interesting and productive ways that influence and inhabitation combined in attempts to create a viable, non-Aristotelian cosmology, but it is also intended to demonstrate that no history of celestial physics in this period can be complete without appreciating the affinity between these two concepts. Accompanying this survey is a closer look at the philosophical systems of Kenelm Digby (1603–1655) and Thomas White (1593–1676), and the refutation of White's cosmology by Thomas Hobbes (1588–1679). This brief analysis will serve to highlight some of the implications of a mechanistic approach to biological and physical forces for theories of celestial influence and inhabitation. It will also serve as an introduction to the second argument of this book—that throughout the seventeenth century certain thinkers began to consciously oppose influence and inhabitation as rival teleological paradigms of astronomical cosmology.

This argument will be progressed in Chap. 6, which will look at the enmity towards astrology expressed by several prominent advocates for pluralism, finishing with a study of the de-astrologising tendencies within the tradition of Newtonian natural theology. The conclusion of this book looks ahead, proposing new ways to consider the scientific and cultural trajectory of pluralism over the last three centuries up until the present day. In the eighteenth century, it will be suggested, pluralism, and the increasing belief in ET life, *replaced* astrology in a certain sense. That is to say, it assumed many of the cosmological, cultural, and psychological functions which were previously performed by astrology. It fulfilled the teleological requirements of the New Astronomy. Put simply, God did not create the planets and the stars in order to influence the earth but rather as abodes for a myriad of other creatures. But more than this, it took astrology's place as the conjectural or speculative side of astronomy. In this sense, it served, and continues to serve, as an impetus for the astronomical sciences and a vehicle for popular engagement with the celestial realm.

The demonstration of these arguments will, it is hoped, contribute to the history of both astrology and pluralism. In the medieval and Renaissance periods astrology was considered by many to be the queen of the sciences, embodying the ultimate utility of mathematical and astronomical studies. The recent special issue of *Early Science and Medicine* on the marginalisation of astrology presents a consensus view, perhaps unsurprisingly, that this is a process not well enough understood (Vermij and Hirai 2017). A step towards a better understanding can be taken by appreciating the appeal of pluralism and how it came to be placed in juxtaposition to astrology. At the same time, the history of our modern obsession with life elsewhere in the universe can be enriched by examining how the new paradigm, to use Kuhnian phrasing, grew out of the old. In effect, this research aims to build a bridge between the 'antiquarianism' of the history of astrology and the 'presentism' of the history of pluralism.[12]

1.2 Two Histories

The history of pluralism, or the ET life debate, was the focus of several monograph studies in the twentieth century. Arthur Lovejoy's famous work *The Great Chain of Being* (1936) used pluralism as a case study to advocate his new methodology for a history of ideas. His main argument was that the driving force behind the change from the medieval to the modern cosmos was not Copernicanism or scientific astronomy, but rather the revival of certain Platonic metaphysical preconceptions (Lovejoy 1948, 99). One such Platonic conception, treated by Lovejoy as a 'unit-idea', was the Principle of Plenitude, which dictated, to put it briefly, that the infinite power and goodness of the creator must be realised in an infinite creation. Whatever can be, is. Alexandre Koyré agreed to some extent with Lovejoy's emphasis on Platonism, but his own famous work, *From the Closed World to the Infinite Universe* (Koyré 1957), placed much more emphasis on Copernicus and the improvement of mathematical astronomy.[13]

Koyré's mild rebuttal was strengthened implicitly by Stephen J. Dick in his 1982 book *Plurality of Worlds: The Origins of the Extraterrestrial Life Debate from Democritus to Kant*, which takes its approach from the history of science and accentuates the role played by observational evidence. This work, insightful and broad in scope, remains effectively the standard text on the pre-modern history of pluralism. Its sequel, chronologically speaking, is Michael Crowe's *The Extraterrestrial Life Debate, 1750–1900: The Idea of a Plurality of Worlds from Kant to Lowell* (1986). He disagreed slightly with Dick, arguing that the huge gaps between observation

[12] In his review of the English translation of Guthke's *Der Mythos der Neuzeit*, David Lux alluded to the dangers of treating history as a 'pursuit of origins', an approach championed by the 'New History' school of the early twentieth century: 'As critics of the New History pointed out all too clearly, an avowed presentism very often yields a telescopic effect, one in which the historian's emphasis on modernity and progress can overpower important subtleties and nuances of historical action' (Lux 1994, 121). The pejoratives 'antiquarian' and 'presentist' are used here only for rhetorical purposes, to highlight the potential benefits of a conjoined and comparative history. There is of course an irony in proposing an early modern history of astrobiology as an antidote to presentism. It is left to the reader to judge whether this is a useful hypocrisy.

[13] Paolo L. Rossi argued that the history of the dispute about inhabited worlds did not coincide with the narratives of either Lovejoy or Koyré, nor with the history of imaginary voyages, but was rather a history *sui generis*. See Rossi 1972, 157. The foundational text for histories of lunar voyage literature is Nicolson 1948.

and belief regarding questions of ET life suggest that religious and philosophical factors, such as teleology and the principle of plenitude, played the primary role. It is unlikely that this disagreement on emphasis can be resolved, as it relies heavily on subjective choices of which historical figures, and which parts of their work, to prioritise in the history of pluralism.

In between the works by Dick and Crowe, the German literary and cultural historian Karl S. Guthke published *Der Mythos der Neuzeit: Das Thema der Mehrheit der Welten in Literatur- und Geistesgeschichte von der Kopernikanischen Wende bis zur Science Fiction* ('The Myth of Modern Times: The Theme of a Plurality of Worlds in Literary and Intellectual History from the Copernican Revolution to Science Fiction', 1983).[14] Guthke's main argument, and the reason for the inclusion of literature, is that the belief in a plurality of inhabited worlds has become, as the title suggests, the predominate 'myth' of modernity—a 'new gospel'(Guthke 1990, 35).[15] He identifies many drives towards pluralism, such as religious thinking, philosophy, analogy, observation, as well as imagination and fiction. His focus on the anthropological dimensions of the belief in pluralism, both in its modern and early modern periods, is a welcome addition to the scientific and philosophical histories.

Guthke's thesis is a reminder of a very important point. The history of the plurality of worlds hypothesis, written in a time when it underpins so much of our cosmology, culture, and scientific endeavour, is accordingly written as the history of a winning idea; a 'truth'. As such it falls victim in some degree to the presentism, bias, and telescopic distortion that such historicism entails, although in fairness, this narrow focus should be forgiven as a consequence of the grand chronological scope of the principal works. There have since been some attempts to contextualise aspects of this long history. An article by Nathaniel Wolloch looks at seventeenth-century philosophers who theorised about ET life and compares this to the same philosophers' theories on animals (Wolloch 2002). These links between pluralist, theriophilic (pro-animal) and anti-theriophilic ideas are valuable, in Wolloch's view, for a better understanding of the issue of anthropocentrism in this period. This book will replicate his methodology to a certain extent for the links between pluralism, astrological and anti-astrological theories, with similar implications for our historical understanding of anthropocentrism.

As for astrology itself, its demise as a scientific discipline in the seventeenth century is usually considered as part of a wider development in Western civilisation often called the 'disenchantment of the world' or the 'decline of magic'. This latter phrase is of course part of the title of Keith Thomas' seminal work, which focused mainly on the historical situation in England, yet has in many ways set the terms of the debate more broadly in the decades since, especially with its integration of scientific, religious, and social factors (Thomas 1971).[16] His invocation of political and

[14] The English translation has an altered title: Karl Siegfried Guthke, *The Last Frontier: Imagining Other Worlds, from the Copernican Revolution to Modern Science Fiction* (Guthke 1990).

[15] This idea is not original to Guthke. See Chernyshova 2004 [1972].

[16] The phrase 'disenchantment of the world' comes from the sociologist Max Weber (1864–1920). For an introduction to the historiography of disenchantment, see Walsham 2008.

social influences has since been developed further, especially by Patrick Curry, who focuses again primarily on England (Curry 1989). The story of astrology's demise in these cases obviously involves the identification of a divide, or at least a continuum, between low and high astrology, because we see that 'popular astrology' in some form has survived up until the present day.

High astrology, which would include what we might think of as 'scientific astrology', was assumed, by traditional histories of science, to have declined simply because it was disproved or discovered to be groundless. At the very least it did not conform to the definition of science as it was being developed in the seventeenth century. For Popper, astrology failed because it was unfalsifiable; for Kuhn, because it didn't fit the new paradigm (Popper 1965, 37; Kuhn 1970, 7). Others focus on the anti-astrological impact of certain key works disseminated in the sixteenth century, such as Giovanni Pico della Mirandola's *Disputationes adversus astrologiam divinatricem* ('Arguments against Divinatory Astrology', 1496), or Copernicus' *De revolutionibus orbium coelestium* ('On the Revolutions of the Celestial Spheres', 1543). This theory must somehow account, however, for the continuation of astrological science well into the seventeenth century by prominent philosophers and scientists, including Copernicans. More recent historians of science and astrology have taken a more multifaceted approach to the problem.[17] Some, such as Curry, have argued that the story of astrology's scientific marginalisation is not complete unless it takes into account how the new science or natural philosophy appropriated what were in fact astrological theories, and then made them 'safe' by a process of renaming and reinterpretation (Curry 1991, 282–85). John Henry has written about the contribution of magic and the occult sciences to the new philosophies, including especially the notion of action at a distance (Henry 2008).[18] This was followed by a 'fragmentation' which saw the appropriation of certain aspects and the abandonment of others into a redefined category of vulgar magic.

There are several more general works which have linked astrology and pluralism to some extent, or discussed them in a similar context.[19] One in particular is worth mentioning here, and that is Peter Harrison's *The Bible, Protestantism, and the Rise of Natural Science* (Harrison 1998). In this work, Harrison discusses the dilemma faced by the advocates of physico-theology who wanted to reconcile the new astronomy, with its increased dimensions and countless new and distant bodies, with the teleological tradition of natural philosophy. He comments that the 'decline of astrological prognostication and the related concept of celestial influences … made the problem more acute' (Harrison 1998, 179). The suggestion is that the invocation of ET life as a teleological principle filled in a gap left by an already marginalised astrological tradition. While this was undoubtedly true in certain instances, it is definitely not the whole story. A closer look at the historical record will reveal that,

[17] See in particular Granada et al. 2016; Vermij and Hirai 2017. Few historians engage with the subject of pluralism in this regard, one exception being Vermij 2016, although he comes to different conclusions.

[18] See also Hutchison 1982.

[19] Some examples are Roos 2001; Westman 2011; Omodeo 2014.

beyond the appearance of coincidence or correlation, there are positive causal connections linking the trajectories of the astrological and pluralist traditions.

That being said, the current discussion will not necessarily argue against any of the studies so far mentioned. Rather, the addition of astrology to the history of pluralism, and vice versa, suggests an extra dimension to the debate which expands on and complements certain existing historical theories. For example, Chapters 2, 3, and 4, detailing the support given to pluralism by Platonic theories of living celestial bodies exchanging mutual and sympathetic influences, can be seen as an extension of Lovejoy's thesis that the transition to a modern worldview was stimulated by a resurgent Platonism. It will add to these, however, a suggestion of the importance of Stoic philosophy. At the same time, some of the cosmological developments analysed by Dick and Crowe can be better understood contextually by reference to the field of astrology, which is under-represented in their studies. Our understanding of novel philosophies from the Renaissance and early modern periods which are pro-pluralism on some level will be enriched by an analysis of theories of celestial influence within those same philosophies. It will be shown that astrological ideas were not simply a casualty of the move from the closed world to the infinite universe, but in fact an active participant.

Building on these foundations, Chap. 5 will hopefully demonstrate the extent of common ground between the histories of astrology and pluralism in the seventeenth century. Early modern celestial physics encapsulated both influence and inhabitation, and so it is no surprise that the two ideas formed part of the same intellectual and cultural milieu. Chapters 6 and 7, which argue that pluralism was put in opposition to, and then replaced astrology, continue the line of research pursued by Guthke, tracing the cultural and anthropological aspects of a belief in pluralism and extraterrestrial life. The argument is really a more specific rephrasing of Guthke. Rather than the 'myth of modern times', this research suggests that pluralism and a belief in ET life are in some ways the 'astrology of modern times'.[20] This thesis can also be considered in Kuhnian terms, as an example of a paradigm shift.[21] Lynn Thorndike argued that, prior to Newton, astrology was a 'generally recognized and accepted … universal natural law', while Dick believes that 'the idea that abundant life exists in the universe is more than another theory or hypothesis; it is sufficiently comprehensive to qualify as a worldview' (Thorndike 1955, 273; Dick 1996, 135). Can we therefore think of a paradigm shift from influence to inhabitation (or, to phrase it more colloquially, from astrology to aliens) occurring in the seventeenth and eighteenth centuries, with its roots in astronomy but branching far afield into society and culture? This thesis, if correct, suggests that there was no disenchantment in the early modern period, at least not astronomically speaking. What we find instead is the *terms* of enchantment metamorphosing in step with a changing cosmology.

[20] You could of course say that astrology was the myth of pre-modern times. For an interesting discussion of myth and science more generally, see Carroll 1980.

[21] The concept of paradigm shift as a way to understand scientific change was developed in Kuhn 1962. The merits of applying the term in this context will be considered in Chap. 7.

Both astrology and pluralism satisfied the teleological requirements of astronomy, but, perhaps more importantly, they both provided what Kuhn would call 'psychological satisfaction' to their respective worldviews (Kuhn 1957, 7). This psychological aspect was in itself a motivating force, providing philosophers with ever more intriguing and pressing questions and problems. When Crowe argues that the search for clues of celestial inhabitation was a primary incentive for the building of large telescopes in the eighteenth century, and that embellishing astronomical works with ideas about extraterrestrial life was a way of interesting a broader public, he demonstrates that pluralism had taken astrology's role as the 'foolish daughter' supporting her wise but poor mother astronomy (Crowe 1986, 555–56).[22] Rienk Vermij has suggested that astrology was abandoned because other scientific and cosmological problems were putting demands on astronomy (Vermij 2014, 171). It was the very real possibility of the existence of ET life, and the plethora of questions and dilemmas that it raised, that placed those demands and took centre stage, where it remains today.

This links to another possible approach: that of Gerard Holton's emphasis on the thematic dimension of science. *Themata* for Holton are 'non-scientific' commitments which can provide the source of an induction or determine the choice or preselection of theories. 'One result of this recognition', argues Holton, 'will be that the dichotomy between scientific and humanistic scholarship, which is undoubted and real at many levels, becomes far less impressive if one looks carefully at the construction of scientific theories' (Holton 1988, 33). Taking heed of Lorraine Daston's lament about the over-historicisation of the history of science (Daston 2016), this book will at least make suggestions for a psychological, sociological and anthropological approach to the history of astronomy. Astrology and pluralism, which bridge the divide between science and speculation, between the professional and the popular, are perfect candidates for such an approach. Stefan Helmreich, in his article on the astrobiological imagination, advocated a cultural historical approach which would aid in 'understanding and uncovering the exuberance of such scientific enterprises as astrobiology, which chase after such overflowing objects as "life"' (Helmreich 2006, 86). The centuries-long exuberance of astrological enterprise is perhaps the best model we have for such an understanding.

This book, while focusing largely on the long seventeenth century, attempts to establish patterns over an even longer time frame. An apology must therefore be made for the inevitable generalisations and cursory treatment of important historical themes and figures. Even though this essay is an attempt, in some degree, to contextualise the history of the ET life debate, the large period covered means that it is just as guilty of a biased selection and prioritisation of sources. To lessen the need for such an apology, it should be made clear that this book is only intended to demonstrate that interactions between astrology and pluralism formed part of the historical trajectories of both traditions. It will be enough to establish that these two particular trends—that questions concerning celestial influence stimulated ideas about possi-

[22] The depiction of astrology as the foolish daughter of astronomy was made by Kepler in his *Tertius interveniens* (1610). See Kepler 1937–, IV, 161.

ble inhabitation, and that the latter contributed to the forces responsible for astrology's decline—were important, but not universal, factors in the history of early modern natural philosophy and cosmology.

1.3 Clarifying Terms

We have already (over-)defined astrobiology. The anachronistic application of this term will be minimised in the body of this work so as not to distract from the attempt to appreciate this history *in situ*. There are several other terms which need clarification before we can proceed. 'Pluralism' is used in this context only as shorthand for the philosophical doctrine that posits a multiplicity of worlds, and so should not be confused with any other sort of pluralism. It should also be noted that this history is concerned almost exclusively with that variety of pluralism which theorises about other earth-like worlds within this one universe—worlds in which generation and corruption occur in a way analogous to the earth. It is not concerned so much with theories of a plurality of *kosmoi* separated by place or time. While pluralism will be the most common and general term employed, discussion will also focus on extraterrestrial (ET) or alien life. Both these expressions are anachronisms, and bring with them associations from modern popular culture and science fiction. Nevertheless, they are retained partly because this book, especially in the later chapters, is interested in exactly those broader cultural manifestations and ramifications of pluralism.

This brings up the question of what does and does not qualify as ET life. Like the modern science of astrobiology, this book will use terrestrial life as the example to be applied to life as a more general concept. That is to say, ET life is considered in the form of plants and animals which are corporeal and which live (and die) on the other celestial globes. This therefore excludes planetary intelligences, planetary souls, angels, demons, daemons, gods, and others of the like. This is not an airtight rule, however, and discussion of these forms of life will be necessary from time to time, and indeed the distinction is quite often blurred in the sources themselves. There was, with notable exceptions, very little discussion in this period of what life on other celestial bodies might look like. We will, however, come across interesting and pertinent ideas related to this, such as the tension between plenitude and the uniformity of nature, and the question of what it actually is to be human in a universe with more than one inhabited planet.

What exactly is meant by the term 'astrology' is quite often debated and confused in the scholarship. Many subdivisions were categorised throughout the classical, medieval and Renaissance periods, and even more have been applied by modern historians. We have already mentioned high and low astrology, which is of course a modern delineation. Isidore of Seville (560–636) distinguished between 'natural astrology' and 'superstitious astrology' in his *Etymologies* (III.27). Superstitious astrology is more accurately referred to as 'judicial astrology', and the division is usually considered to separate the influence of the heavens over the natural world from its influence over humans and human events, which impinged on free will. A distinction can also be made between practical astrology and different levels of astrological theory, that is, theory oriented towards practice and theory concerning celestial influence as a subject of natural philosophy.[23] It is this last meaning which is the focus of the current discussion. Lemay called it theoretical, learned astrology. Others, including Darrel Rutkin in his PhD dissertation, have called it 'scientific astrology' (Lemay 1987, 60–63; Rutkin 2002, 21–22).[24] Rutkin has since switched his emphasis onto an 'astrologizing Aristotelian natural philosophy' (Rutkin 2006, 2015, 2018). To this list might be added, as we will see, an 'astrologizing Platonic/ Stoic natural philosophy'. For the sake of simplicity and brevity, however, this book will often use the term 'astrology' by itself, on the understanding that it does not refer, in general, to practice or practical theory.

In the light of such possible confusions, both terminological and philosophical, one can see the attraction of Lovejoy's unit-ideas. If we allow ourselves, with an appropriate awareness of the limits of applicability, to use the same method, we could isolate two such unit-ideas for use in this book: the Principle of Influence and the Principle of Inhabitation. The first principle would be the reduction of astrology to two basic questions: Are the celestial bodies causes? And if so, what effects do they have? The latter principle would be the notion that, simply put, there are mortal living beings on other celestial bodies. A comparative history of these notions, beliefs, principles, unit-ideas, paradigms, *endoxai*, however we might like to think of them, is what is attempted here. It begins with an agonisingly brief description of the history of each from the classical period to the end of the sixteenth century, and with a slightly closer look at some examples of contact between them. The first such contact is made by Plutarch.

References

Aristotle. 1882. *On the Parts of Animals*. Tran. William Ogle. London: Kegan Paul.

Berthelot, René. 1938. *La pensée de l'Asie et l'astrobiologie*. Paris: Payot.

Carroll, Michael P. 1980. Le pensée scientifique: Myth and the popularity of scientific theories. *Structuralist review* 2: 49–58.

Catling, David C. 2013. *Astrobiology: A very short introduction*. Oxford: Oxford University Press.

Chalmers, Alan. 2016. Viewing past science from the point of view of present science, thereby illuminating both: Philosophy versus experiment in the work of Robert Boyle. *Studies in History and Philosophy of Science Part A* 55: 27–35.

Chernyshova, Tatiana. 2004. Science fiction and myth creation in our age. In *Science fiction studies*, vol. 31, 345–357. [Translated by Istvan Csicsery-Ronay, Jr., from the original: Chernyshova, Tatiana 1972. Nauchnaya fantastika i sovremennoe mifotvorchestvo. In *Fantastika-72*, ed. E. Brandis and V. Dmitrevsky, 288–30. Moscow: Molodaya gvardia.].

Cockell, Charles S. 2015. *Astrobiology: Understanding life in the universe*. Chichester: Wiley.

[23] For a discussion of the terminology, see Rutkin 2002, 20–22.

[24] See also North 1986; Lindberg 1992, 74–90; Grant 1994, chap. 19.

Crowe, Michael J. 1986. *The extraterrestrial life debate, 1750–1900: The idea of a plurality of worlds from Kant to Lowell*. Cambridge: Cambridge University Press.
Curry, Patrick. 1989. *Prophecy and power: Astrology in early modern England*. Princeton: Princeton University Press.
———. 1991. Astrology in early modern England: The making of a vulgar knowledge. In *Science, culture and popular belief in renaissance Europe*, ed. Stephen Pumfrey, Maurice Slawinski, and Paolo L. Rossi, 274–291. Manchester: Manchester University Press.
Daston, Lorraine. 2016. History of science without structure. In *Kuhn's* Structure of scientific revolutions *at fifty: Reflections on a science classic*, ed. Robert J. Richards and Lorraine Daston, 115–132. Chicago/London: University of Chicago Press.
Demets, René. 2012. Darwin's contribution to the development of the panspermia theory. *Astrobiology* 12: 946–950.
Dick, Steven J. 1982. *Plurality of worlds: The origins of the extraterrestrial life debate from Democritus to Kant*. Cambridge/New York: Cambridge University Press.
———. 1996. Other worlds: The cultural significance of the extraterrestrial life debate. *Leonardo* 29: 133–137.
Dick, Steven J., and James Edgar Strick. 2005. *The living universe: NASA and the development of astrobiology*. New Brunswick: Rutgers University Press.
Dunér, David, Christophe Malaterre, and Wolf Geppert. 2016. The history and philosophy of the origin of life. *International Journal of Astrobiology* 15.
Dunér, David, Erik Persson, and Gustav Holmberg. 2013. *The history and philosophy of astrobiology: Perspectives on extraterrestrial life and the human mind*. Cambridge: Cambridge Scholars Publishing.
Granada, Miguel Á., Dario Tessicini, and Patrick J. Boner, eds. 2016. *Unifying heaven and earth: Essays in the history of early modern cosmology*. Barcelona: Edicions Universitat Barcelona.
Grant, Edward. 1994. *Planets, stars, and orbs: The medieval cosmos*, 1200–1687. Cambridge: Cambridge University Press.
Guthke, Karl Siegfried. 1983. *Der Mythos der Neuzeit: das Thema der Mehrheit der Welten in der Literatur- und Geistesgeschichte von der kopernikanischen Wende bis zur Science Fiction*. Bern: Francke.
———. 1990. *The Last Frontier: Imagining Other Worlds, from the Copernican Revolution to Modern Science Fiction*. Trans. Helen Atkins. Ithaca/London: Cornell University Press.
Harrison, Peter. 1998. *The Bible, Protestantism, and the rise of natural science*. Cambridge/New York: Cambridge University Press.
Helmreich, Stefan. 2006. The signature of life: Designing the astrobiological imagination. *Grey room* 23: 66–95.
Henry, John. 2008. The fragmentation of renaissance occultism and the decline of magic. *History of Science* 46: 1–48.
Holton, Gerald. 1988 [1973]. *Thematic origins of scientific thought: Kepler to Einstein*, Rev. ed. Cambridge, MA/London: Harvard University Press.
Hutchison, Keith. 1982. What happened to occult qualities in the scientific revolution? *Isis* 73: 233–253.
Kepler, Johannes. 1937–. *Gesammelte werke*. Edited under the supervision of the Bayerische Akademie der Wissenschaften, 25 vols published so far. Munich: C.H. Beck.
Koyré, Alexandre. 1957. *From the closed world to the infinite universe*. Baltimore: Johns Hopkins Press.
Kuhn, Thomas S. 1957. *The Copernican revolution: Planetary astronomy in the development of Western thought*. Cambridge, MA: Harvard University Press.
———. 1962. *The structure of scientific revolutions*. Chicago: University of Chicago Press.
———. 1970. Logic of discovery or psychology of research? In *Criticism and the growth of knowledge*, ed. Imre Lakatos and Alan Musgrave, 1–24. Cambridge: Cambridge University Press.
Lagrésille, Henry. 1902. *Le fonctionnisme universel, essai de synthèse philosophique, Monde sensible*. Paris: Fischbacher.

Lemay, Richard. 1987. The true place of astrology in medieval science and philosophy: Towards a definition. In *Astrology, science, and society: Historical essays*, ed. Patrick Curry, 57–73. Woodbridge/Wolfeboro: Boydell Press.

Leunissen, Mariska E.M.P.J. 2009. Why stars have no feet: Explanation and teleology in Aristotle's cosmology. In *New perspectives on Aristotle's* De caelo, ed. Alan C. Bowen and Christian Wildberg, 215–237. Leiden: Brill.

Lindberg, David C. 1992. *The beginnings of Western science: The European scientific tradition in philosophical, religious, and institutional context, 600 B.C. to A.D. 1450*. Chicago/London: University of Chicago Press.

Lovejoy, Arthur O. 1948 [1936. *The great chain of being: A study of the history of an idea*. Cambridge, MA: Harvard University Press.

Lux, David S. 1994. Review of *The last frontier: Imagining other worlds, from the Copernican revolution to modern science fiction. The Journal of Modern History* 66: 120–121.

Michel, Paul-Henri. 1973. *The Cosmology of Giordano Bruno*. Trans. R.E.W. Maddison. London: Methuen.

Morrison, David. 1903. New books. *Mind* 12: 266–267.

Nicolson, Marjorie Hope. 1948. *Voyages to the moon*. New York: Macmillan.

North, John D. 1986. Celestial influence: The major premise of astrology. In *"Astrologi hallucinati": Stars and the end of the world in Luther's time*, ed. Paola Zambelli, 45–100. Berlin/New York: W. de Gruyter.

Omodeo, Pietro Daniel. 2014. *Copernicus in the cultural debates of the Renaissance: Reception, legacy, transformation*. Leiden: Brill.

Philosophical Periodicals. 1934. *Mind* 43: 266–274.

Popper, Karl R. 1965 [1963]. *Conjectures and refutations: The growth of scientific knowledge*, 2nd Rev ed. London: Routledge and K. Paul.

Roos, Anna Marie Eleanor. 2001. *Luminaries in the natural world: The sun and the moon in England, 1400–1720*. New York/Oxford: Peter Lang.

Rossi, Paolo L. 1972. Nobility of man and plurality of worlds. In *Science, medicine, and society in the Renaissance: Essays to honor Walter Pagel*, ed. Allen G. Debus, 131–162. New York: Science History Publications.

Rutkin, H. Darrel. 2002. Astrology, natural philosophy, and the history of science, c. 1250–1700: Studies toward an interpretation of Giovanni Pico della Mirandola's *Disputationes adversus astrologiam divinatricem*. PhD dissertation, University of Indiana.

———. 2006. Astrology. In *The Cambridge history of science. Vol. 3, Early modern science*, ed. Katharine Park and Lorraine Daston, 541–561. Cambridge: Cambridge University Press.

———. 2015. Understanding the history of astrology accurately: Methodological reflections on terminology and anachronism. *Philosophical Readings* 7: 42–54.

———. 2018. How to accurately account for astrology's marginalization in the history of science and culture: The central importance of an interpretive framework. *Early Science and Medicine* 23: 217–243.

Sandoz, Raphaël. 2016. Whewell on the classification of the sciences. *Studies in History and Philosophy of Science Part A* 60: 48–54.

Santos, Charles Morphy D., Leticia P. Alabi, Amâncio C.S. Friaça, and Douglas Galante. 2016. On the parallels between cosmology and astrobiology: A transdisciplinary approach to the search for extraterrestrial life. *International Journal of Astrobiology* 15: 251–260.

Teed, Cyrus. 1908. The science of astro-biology. *Flaming sword* 22.1, January 15: 6–7.

Temple, Robert. 2007. The prehistory of panspermia: Astrophysical or metaphysical? *International Journal of Astrobiology* 6: 169–180.

Thomas, Keith. 1971. *Religion and the decline of magic: Studies in popular beliefs in sixteenth and seventeenth century England*. London: Weidenfeld & Nicolson.

Thorndike, Lynn. 1955. The true place of astrology in the history of science. *Isis* 46: 273–278.

Vermij, Rienk. 2014. The marginalization of astrology among Dutch astronomers in the first half of the 17th century. *History of Science* 52: 153–177.

———. 2016. Seventeenth-century Dutch natural philosophers on celestial influence. In *Unifying heaven and earth: Essays in the history of early modern cosmology*, ed. Miguel Á. Granada et al., 291–315. Barcelona: Edicions Universitat Barcelona.

Vermij, Rienk, and Hiro Hirai, ed. 2017. Marginalisation of astrology. Special issue. *Early Science and Medicine* 22.

Walsham, Alexandra. 2008. The reformation and "the disenchantment of the world" reassessed. *The Historical Journal* 51: 497–528.

Westman, Robert S. 2011. *The Copernican question: Prognostication, skepticism, and celestial order*. Berkeley/London: University of California Press.

Whitfield, Peter. 2001. *Astrology: A history*. London: British Library.

Wolloch, Nathaniel. 2002. Animals, extraterrestrial life and anthropocentrism in the seventeenth century. *The Seventeenth Century* 17: 235–253.

Chapter 2
Celestial Influence as an Aid to Pluralism from Antiquity to the Renaissance

Certain people have said that there are as many species of things in the earth as there are stars. If the earth thus contracts the influence of all the stars into individual species, why does it not happen similarly in the regions of other stars, which receive the influences of the others?

Nicholas of Cusa

Abstract This chapter begins in the classical period, discussing Plutarch's *On the Face which Appears in the Orb of the Moon* as an example of how closely connected theories of influence and inhabitation were when considering the nature of the moon as a potential 'other earth'. The importance of teleology in the discussion is highlighted, as is the prominence of Platonic and Stoic cosmological elements. The chapter then proceeds to the medieval period and looks at the development of an 'astrologizing Aristotelian natural philosophy' and in particular its relevance in two areas: the question of natural generation and the question of divine providence. This astrological natural philosophy was the dominant cosmological paradigm against which (but also within which) the plurality of worlds tradition would eventually develop. The chapter then moves on to Nicholas of Cusa, whose work *On Learned Ignorance* demonstrates how celestial influence was connected to the question of pluralism. After Cusa, the chapter looks at other philosophers who introduced novelties into celestial natural philosophy, such as Marcellus Palingenius, Thomas Digges, Giambattista Benedetti, Francesco Patrizi and Giordano Bruno. In each case the question of celestial inhabitation is shown to be closely linked to questions of influence.

Cusa 1932, 109: 'Dixerunt quidam tot esse rerum species in terra, quot sunt stellae. Si igitur terra omnium stellarum influentiam ita ad singulares species contrahit, quare similiter non fit in regionibus aliarum stellarum influentias aliarum recipientium?'

© Springer Nature Switzerland AG 2019

J. E. Christie, *From Influence to Inhabitation*, International Archives of the History of Ideas Archives internationales d'histoire des idées 228,
https://doi.org/10.1007/978-3-030-22169-0_2

Keywords Astrology · Extraterrestrial life · Plutarch · Nicholas of Cusa · Aristotelianism · Giambattista Benedetti · Francesco Patrizi · Giordano Bruno · Astrobiology

2.1 Plutarch and Classical Pluralism

If one is writing a history of the plurality of worlds tradition in Western thought, the natural place to begin is with the Greek atomist school of Democritus and Epicurus, and their Latin champion Lucretius (c.99–c.55 BCE). The philosophy of these atomists, however, lends itself more to a belief in a multiplicity of more-or-less self-contained *kosmoi*. That is, the infinity of space and matter suggested the reproduction, in another place, of an inhabited earth surrounded by planets and stars. This chapter, on the other hand, is more concerned with that particular form of pluralism which asks whether the other celestial bodies could be worlds like the earth. Atomism would await reinvention by the likes of Bruno, Gassendi and Descartes before it encompassed this latter form of pluralism. More relevant, although far less penetrable, are the philosophical contributions of certain pre-Socratics, such as Thales of Miletus, and the Pythagoreans, who argued for the earth-like nature of the moon.[1] Fragments from the writings of these philosophers describe the moon as material in a sense similar to the earth, furnished with mountains, valleys, seas, and possibly even plants, animals and inhabitants. They were preserved in sources such as Pseudo-Plutarch and Stobaeus (fl. fifth-century CE), whence they found their way into the Latin tradition.[2]

If one is writing a history of Western astrology, on the other hand, the atomists play little to no part.[3] That story begins with the adaptation of Near-Eastern astrological practices to Platonic astral theology, Stoic determinism, and Aristotelian physics.[4] In Aristotle's cosmology, the *primum mobile* transferred motion down through the nested spheres of the planets and into the mutable sublunar world, while the friction caused by that motion produced a vital heat, proceeding primarily from the sun. This physical form of top-down celestial influence was juxtaposed in the classical tradition by Stoic cosmic sympathy and Platonic forms, which created a relationship between the microcosm and the macrocosm that was at the same time both symbolic and tangible. Even Aristotle's natural philosophy, however, contained a more direct connection between man and the cosmos at large. In his work, *On the Generation of Animals*, he stated that the physical substance which carries the soul, *pneuma*, was 'analogous to the element which belongs to the stars', i.e. ether (*GA*, II.3, 736b38–39; Aristotle 1943, 171). There is then an analogy between this kind of

---

[1] For a discussion of the Greek sources for the pluralist philosophy, see McColley 1936, 385–92.

[2] Pseudo-Plutarch, *Placita philosophorum*, II.25–30; Stobaeus, *Eclogues*, I.17–25.

[3] Consider this quote from Jaki 1978, 10: 'If there was a saving grace in the cosmological dicta of Epicurus and Lucretius, it consisted in the absence of fantasy and astrology.'

[4] A good summary of this process is given in von Stuckrad 2016.

pneuma and that believed by the Stoics to permeate the entire universe, acting as the cohesive agent and indeed as the soul of both the world itself and everything in it.

The celestial science of Plato and the Stoics was highly teleological and indeed theological, an approach which was attacked by Epicurus and Lucretius, leading to the conclusion that these atomists had no time for astrology (Warren 2004, 355; Campion 2008, 179). Yet, while a focus on materialistic atheism and the desire to free people from superstition does seem to align atomism with an anti-astrological stance, we need only recall a few passages from Lucretius to know that his view of nature was not free from celestial causation:

> Then be it ours with steady mind to clasp
> The purport of the skies- the law behind
> The wandering courses of the sun and moon;
> To scan the powers that speed all life below...[5]

This passage suggests a general effect of celestial influence similar to that found in Aristotle. Platonic elements arise, too, as when we read that 'we all from seed celestial spring', and further: 'what was sent from shores of ether, that, returning home, the vaults of sky receive'.[6] It is true, however, that an explication, or even refutation, of astrology was not one of Lucretius' primary concerns. This point is made more pertinent when we consider another cosmological work in Latin verse, the heavily-astrological *Astronomica* by Manilius (c.first century CE). Katharina Volk has argued that while Manilius was keen to emulate much of Lucretius' style and content, the work's presentation of 'an orderly cosmos ruled by fate' (i.e. ruled astrologically) was a direct attack on the chaotic and random world of the atomists (Volk 2009, 192).[7] This use of astrology as an instrument and demonstration of cosmic order will become increasingly relevant later in our discussion.

In terms of classical philosophy, then, the relevant starting point for this study is Plutarch (c.46–120 CE), and in particular his *De facie quae in orbe lunae apparet* ('On the Face Which Appears in the Orb of the Moon').[8] The interlocutors in this incomplete dialogue discuss the opinions of the various philosophical schools concerning the nature and purpose of the moon, dealing not only with its essence and potential habitability, but also with issues to do with its motion, gravity and influence. More so than the fragments related above, Plutarch's treatise reveals the depth of interest in the nature of our nearest celestial neighbour in the classical period (Dick 1982, 20). The lunar theories contained therein foreshadow later innovations such as those of Cusa, Copernicus and others. One of its central questions is whether we can consider the moon to be another earth. Several objections could be raised to

---

[5] Lucretius, *De rerum natura*, I, 127–30: '...qua propter bene cum superis de rebus habenda / nobis est ratio, solis lunaeque meatus / qua fiant ratione, et qua vi quaeque / gerantur in terris.' English verse from Lucretius 2004, 4.

[6] Lucretius, *De rerum natura*, II, 991: 'Denique caelesti sumus omnes semine oriundi...'; ibid., II, 1000–1: '...quod missumst ex aetheris oris, / id rursum caeli rellatum templa receptant.' Lucretius 2004, 64.

[7] See also Hardy 2015, 208; Porter 2016, 486–88.

[8] Quotations are from Plutarch 1957.

this idea. There was the argument, for example, attributed here to the Stoics, that any earth-like body would have a tendency toward the middle of the universe, i.e. the earth. In response, the speaker Lamprias argues that the inclination of falling bodies 'proves not that the [earth] is in the centre of the cosmos, but that those bodies, which when thrust away from the earth fall back to her again, have some affinity and cohesion with her' (Plutarch 1957, 69).[9] It follows that other bodies such as the sun and the moon have the same ability to call things of their nature back to them, and thus to retain completeness in their relative position (Plutarch 1957, 71).

Plutarch's leaning towards Platonic philosophy further supported this assertion, as when Lamprias describes the universe as a living creature that may easily contain all the elements spread throughout, organised according to a rational order (Plutarch 1957, 91–93). On top of the subject of material elements existing in the heavens, there are more objections to be faced concerning the motions of those elements if taken out of the terrestrial realm. So, opposed to an Aristotelian physics which asserted rectilinear motion in the sublunary and circular motion in the superlunary worlds, Plutarch emphasised a Platonic theory of motion that could apply to the universe as a whole.[10] In the beginning, we learn, all matter was separate, with each unit performing its own motions, 'scornful' of the others. This was until Providence introduced desire, affinity and friendship into nature. At this point, motions stopped being 'natural' in that word's prior sense, and things began to change place according to what was 'better', producing harmony and communion through mutual intercourse (Plutarch 1957, 83–85). He stated it thus:

> … no part of a whole all by itself seems to have any order, position, or motion of its own which could be called unconditionally 'natural.' On the contrary, each and every such part, whenever its motion is usefully and properly accommodated to that for the sake of which the part has come to be and which is the purpose of its growth or production, and whenever it acts or is affected or disposed so that it contributes to the preservation or beauty or function of that thing, then, I believe, it has its 'natural' position and motion and disposition (Plutarch 1957, 89–91).

When this rationale for motion is applied to the heavenly bodies, it creates a system which is somewhere between inanimate physics and planetary intelligences. Motions should be understood teleologically, in terms of purpose and intention, or even organically, if the universe be thought of as a living being—yet they are definitely ordered according to a notion of desire that is providentially ordained.

Concerning astrology, Nicholas Campion makes the point that it was Plutarch, rather than his more influential (in this field) contemporaries Vettius Valens and Ptolemy, who articulated the moon's planetary qualities, pointing to this passage from *Isis and Osiris*:

---

[9] Knox argues that Copernicus probably didn't read *De facie*. However, Knox mentions two other works by Plutarch which repeat this theory of gravity: Plutarch, *De Stoicorum repugnantiis*, XLIV, 1054E–55B; *De defectu oraculorum*, XXVI, 424E. See Knox 2005, 182–89.

[10] This does conflict, however, with Plutarch's cosmological dualism, which contrasted the changeable sublunary realm with the regular celestial realm, as stated in *Isis and Osiris*, 369b–d. See Chlup 2000, 150.

> The moon, because it has light that is generative and productive of moisture, is kindly towards the young of animals and the burgeoning plants, whereas the sun by its untempered and pitiless heat, makes all growing and flourishing vegetables hot and parched, and through its blazing light, renders a large part of the earth uninhabitable. In fact the actions of the moon are like the actions of reason and perfect wisdom, whereas those of the sun are like beatings through violence and brute strength (Plutarch 1936, 101).[11]

The astrological benefits of the moon become important in *De facie* when the figure of Theon asks to hear about whether the moon is habitable. If it is not, he suggests, then the argument put forward so far that the moon is another earth would lose credibility, as the purpose of the earth is to provide an abode for plants, animals and humans (Plutarch 1957, 157). The first response offered by Lamprias is that the moon being uninhabited would not mean it has no purpose, as the earth has many uninhabited areas (such as deserts, oceans and glaciers) that are nevertheless helpful to living things (e.g. through ocean breezes and the temperate effects derived from melting snow) (Plutarch 1957, 165).[12] This passage could be seen as a response to the anti-teleological stance of the Epicureans, who used deserts as an argument against the design of the cosmos for the benefit of man.[13]

One way that a barren moon still retains purpose concerns its influence. It does this by 'providing reflections for the light that is diffused about her, and for the rays of the stars a point of confluence in herself' (Plutarch 1957, 167). Thus the moon can aid through the qualities that Plutarch had outlined above, with its generative and humidifying light which acts to temper the otherwise scorching rays of the sun.[14] This seemingly obvious effect of moisture from the moon, however, is evidence in itself that it may be habitable. It demonstrates that Theon's arguments of the intense heat and dryness of the moon, based on its long exposure to the sun, cannot hold true, as the effects of moisture must follow from the nature of the body itself. Even if the moon is hot and without rains, this present moisture would be enough for the growth of plants, and Lamprias gives as an example those plants which grow in arid regions of the earth, relishing drought and dying from too much moisture (Plutarch 1957, 171–173).[15] Here we see how the teleological response to the Epicurean 'desert argument' could involve an invocation of a less anthropocentric design argument, which celebrated life in all its forms.

With these arguments, Lamprias proves that those philosophers who think that the moon is made of fire are mistaken, as are those who think that animals on the moon would require everything that they have here on earth. Nature is not everywhere equal, and there are greater differences between animals, he thinks, then

[11] See Campion 2008, 218–19.

[12] This extension of terrestrial deserts to a cosmic scale would be later echoed by Kant. See Kant 1981.

[13] See Coones 1983, 366. On Plutarch's negative attitude towards atomist principles, see Hershbell 1982.

[14] The idea of the moon as a 'point of confluence' that relays other celestial influences to the earth will be echoed by later thinkers. See Sects. 2.3, 2.5 and 3.3.

[15] Plutarch's source is probably Theophrastus, *Historia plantarum* (IV, 3.7; VIII, 1.1, 1.4, 2.6, 3.2, 6.6) and *De causis plantarum* (II, 1.2–4; III, 1.3–6).

between animals and the inanimate world (Plutarch 1957, 175–77). All this prepares him to build a conjecture about the men who might live in the moon:

> It is plausible that the men on the moon, if they do exist, are slight of body and capable of being nourished by whatever comes their way. After all, they say that the moon herself, like the sun which is an animate being of fire many times as large as the earth, is nourished by the moisture on the earth, as are the rest of the stars too, though they are countless; so light and frugal of requirements do they conceive the creatures to be that inhabit the upper region (Plutarch 1957, 177–179).

So, the men on the moon are capable of living off more frugal fare than we are here on earth. But the most crucial aspect of this passage is Lamprias' final argument for the habitation of the moon: that the earth has a reciprocal effect of moisture on the lunar body.

Indeed, he argues that the sun and all the other stars also partake of this earthly, vaporous influence, which sustains each of them as a living being. As much as Plutarch, in this and other works, attempted to discredit Stoicism, this concept is one they had in common. Cicero, in *De natura deorum* (45 BCE), had described the heavenly bodies as animate and divine, and quoted Cleanthes that the sun's fiery nature required nutriment from the moisture of the earth (II.15; Cicero 1896, 50). This moisture is drawn in the form of vapour from soil and water. The sun, the stars, and indeed the entire body of the ether, 'after being fed and renewed by these, pour the same back and draw them again from the same source' (II.46; Cicero 1896, 69). The dialogue also outlined the astrological powers of the planets and luminaries, seeing them as a clear demonstration of the different parts of the universe working together for the wellbeing of the whole (II.46; Cicero 1896, 69).

In Plutarch's *De facie* we see how Platonic and Stoic ideas about the nature of and relationship between the celestial bodies could build a bridge between astrology and cosmological pluralism. The belief that these bodies were living creatures allowed them to be given a passive as well as active role in the continuing harmony of creation—a cosmological trend continued by later Platonists such as Plotinus and Proclus. Plutarch already took the next step from celestial bodies receiving influence to sustain themselves to those same bodies receiving influence to support inhabitants. In this 'curious' work, the projection of life into the larger universe was supported by the reciprocation and mutual interchange of beneficial influences—that is, a laterally-expanded alternative to the limited top-down causal chain of the heavens found in Aristotle.

2.2 The Astrologisation of Generation and Providence

As it was Aristotelian philosophy, however, that dominated Western science up until the seventeenth century, it is worth spending some time at this early stage investigating the development of an astrologised Aristotelianism within medieval and Renaissance natural philosophy and theology. There are two dimensions of this broad subject which are particularly relevant to the relationship between astrology and pluralism. The first is the role of celestial influence in generation, especially of the spontaneous kind; the second is the use of astrology as a way to comprehend God's providence. When large swathes of Greek philosophy were reintroduced into Latin-speaking Europe via the Arabic translation movement beginning in the late eleventh century, they did not come unattached. The commentaries of Ibn Rushd (1126–1198), known to the Latins as Averroes, greatly influenced the interpretation of Aristotle's scientific works, and together with the works on optics and ray theory by al-Kindi (c.801–873) and astrological works such as those of Mashallah (c.740–815), they greatly expanded the scope and importance of celestial influence beyond that found in the works of the Stagirite himself.[16]

This was complemented by a similar astrological bent in the field of medicine. The works of Ibn Sīnā (Avicenna, c.980–1037), which attempted to reconcile the medicine of Hippocrates and Galen with Aristotelian natural philosophy, became standard textbooks up until the eighteenth century. It was, arguably, the practice of medicine that most fully integrated astrology into both natural philosophy and society. It was a concrete manifestation of the philosophical affinity between the life sciences and the science of the stars, and it was also the last scientific discipline to jettison astrological tenets from its curriculum.[17] The theory of critical days linked the phases of the moon to the regulation of the four bodily humours.[18] Natal horoscopic astrology was premised on the influence of stellar and planetary configurations on the body and mind of a new born.[19] Doctors also asked astrological 'questions', basing their diagnoses and prognoses on a horoscope cast at the time of consultation.[20] What interests us here, however, are more general questions of the role of celestial influence in generation.

Aristotle was somewhat limiting in his theories of celestial influence. He considered the heat transmitted by the motion of the celestial spheres, especially the sun, as a general but not specific cause in the process of generation. It is in this more general sense that we should probably interpret his famous saying that man is begotten by man and the sun (*Physics*, II.2, 194b13). His writings on the subject were more suggestive than definitive, as for example in the analogy between *pneuma* and the ether mentioned above. The stress here should be placed on 'analogy', as his general position emphasised the disconnect between the ethereal and elemental realms (Freudenthal 2002, 111). There was also Aristotle's assertion that 'in all *pneuma* soul-heat is present, so that in a way all things are full of Soul' (*GA*, III.11, 762a18–27; Aristotle 1943, 357). When these general suggestions were expanded upon by commentators such as Alexander of Aphrodisias (fl. 200 CE) and later Arabic philosophers, it was Platonic and Stoic elements that filled out the picture

[16] It should be noted that works on astrology and magic were those most eagerly searched for and translated by the Latin translators. See Burnett 2009.

[17] For a comprehensive study of Avicenna's synthesis of cosmology and medicine, see Jouanna 2001.

[18] On the tradition of critical days, see Pennuto 2008.

[19] On this topic, see the various contributions in Oestmann et al. 2006.

[20] For an example of this practice, see Kassell 2005.

(Freudenthal 2009). The idea, common to Platonism and Stoicism, that the celestial bodies were living and ensouled creatures, changed the interpretation of their influence. Avicenna believed that they affected the terrestrial realm both through their natural and *volitional* powers (Saif 2015, 75). To the 'physical' dimension, therefore, was added the theory of the emanation of forms, which had been suggested by Themistius (317–c.390) in his commentary on Aristotle (Bertolacci 2013, 46).

The question of heavenly intervention became especially important when philosophers and commentators on Aristotle came to the question of spontaneous generation. This was a widely recognised phenomenon, where living beings were deemed to be generated without seed or copulation. The usual examples were worms from decaying bodies, frogs from mud, and other such things.[21] Aristotle had said little on the subject. His theory involved some sort of material, along with air and its vital heat, being enclosed in something and then forming a 'frothy bubble', which in turn developed into a living organism (Lennox 1982; Gotthelf 2012, 145–49). The type of organism generated was dependent both on the material and the place or situation in which it was enclosed (*GA*, III.11, 762a18–27). In comparison to the usual method of generation, the material was the female principle, while the male or motive principle was supplied by the vital or soul-heat.

Avicenna found this explanation insufficient, and added to this process the action of both the heavenly bodies and the so-called Active Intellect (Kruk 1990, 273–74; Bertolacci 2013, 42). These were responsible for the imposition of form onto the suitable and harmonious mixture of the elements. Remke Kruk has linked Avicenna's theory of spontaneous generation to the 'chain of being', citing the idea that there was no clear demarcation between the realms of nature, and that there was a gradual improvement in living creatures connected to the supposed hierarchy of beings (Kruk 1990, 276). Thus, the elements could turn into plants, plants into animals, etc. In the philosophy of Averroes, we see the role of the heavenly bodies obtaining even more prominence in the wake of his rejection of Avicenna's reliance on the Active Intellect (Hirai 2011, 39; Bertolacci 2013).[22] In the Commentator's treatment of spontaneous generation, we clearly see how Platonic his interpretation of Aristotle was. In his *Epitome of Metaphysics*, he asserted that in cases of generation without a seed, there was need of a 'principle from without', and explained that 'the ultimate mover is, in Aristotle's system, the celestial bodies though the mediation of soul-powers emanating from them'.[23]

Later in the same work he made a clearer connection between the life of the celestial bodies and the life they engendered on earth:

> …some of the blended substances are ensouled owing to the heavenly bodies. This is why Aristotle says that man is generated by man and the sun. The reason for his holding this

[21] While it may now seem bizarre, the idea of spontaneous generation is central to the history of astrobiology as a precursor to abiogenesis—the idea that, under the right circumstances, inanimate matter can give rise to living organisms. On the subject of astrobiology's engagement with questions about the origin of life, see Dick and Strick 2005, 224–26; Świeżyński 2016.

[22] See also Averroes, *Long Commentary on Aristotle's Metaphysics*, VII.31.

[23] Averroes, *Epitome of the Metaphysics*, Section 28, translated in Davidson 1992, 239.

> view is that a man is generated by a man like himself and that because those [celestial] bodies are alive they can endow with life what is [down] here [in the sublunary world]. For only a body whose nature is to be ensouled can move matter to the animate [i.e. soul-] perfection.[24]

The writings of Avicenna and Averroes, therefore, skewed the interpretation of Aristotle's theory of generation in the direction of astrology. This is what Gad Freudenthal describes as the 'astrologization' of Aristotle's cosmos, and is part therefore of the formation of Darrel Rutkin's 'astrologizing Aristotelian natural philosophy' (Rutkin 2002, 20–22; Freudenthal 2009). It was the addition of Platonic, Stoic and also magical dimensions to the question of celestial influence that made Arabic Aristotelianism so attractive to Latin philosophers in terms of its explanatory and practical powers, and at the same time so problematic and controversial.

The celestial science of Mashallah, al-Kindi, Avicenna and Averroes was highly influential on such giants of medieval science and philosophy as Robert Grosseteste (c. 1175–1253), Albertus Magnus (c. 1200–1280), Thomas Aquinas (1225–1274) and Roger Bacon (c. 1219–c. 1292).[25] Al-Kindi, who maintained that terrestrial diversity was the result of the diversity of matter coupled with the varied action of stellar rays (Al-Kindi 1974, 220–21), was a crucial source for Grosseteste and Bacon in their development of theories on optics, light metaphysics, and the 'multiplication of species'. Albertus Magnus was influenced by Avicenna's Platonic cosmology of emanation, in which the One proceeds to the earth through the intermediary stages of the planetary spheres (Campion 2009, 47). He was deeply interested in astrology, and in its role in generation—an interest shared by his student Thomas Aquinas (Rutkin 2013). For Thomas, the heavenly bodies cooperate in the creation of perfect animals, such as men, but cannot be the sole cause, as Avicenna had thought. In the case of certain imperfect animals, however, the power of the celestial bodies is a sufficient cause for the informing of the substance.[26]

As Rutkin has argued, the efforts of figures such as Albertus Magnus and Thomas Aquinas fashioned astrology into an 'integrated theological and scientific system' (Rutkin 2018, 220–21). The ideas of these medieval natural philosophers maintained a strong influence throughout the Renaissance and early modern periods. Astrology was taught at many of Europe's largest universities, such as Padua, Bologna and Paris, as an integral part of mathematics, natural philosophy and medicine. In this last field, the programme of astrological medicine developed by Pietro d'Abano (1257–c. 1316) was particularly important (Siraisi 1973). Brian Copenhaver has demonstrated the direct impact of scholastic astrology on the magic and philosophy of Marsilio Ficino (1433–1499) (Copenhaver 1984, 524). On the issue of spontaneous generation, Ficino took Aristotle's theory in a different Platonic direc-

[24] Averroes, *Epitome of the Metaphysics*, Section 65, translated in Davidson 1992, 161. See also Freudenthal 2002, 120–21.

[25] See Saif 2015, chap. 4.

[26] Thomas Aquinas, *Summa theologiae*, I, q. 91, a. 2. For further scholastic opinions about the role of the heavens in generation, see the discussion of the question 'Can celestial bodies generate living things?' in Grant 1994, 579–86.

tion. Averroes had quoted Themistius, who had linked Aristotle's *pneuma* or soul-heat to the idea of a World-Soul. Ficino, inspired at the same time by Plotinus' *Enneads* and Proclus' commentary on Plato's *Timaeus*, took this popular idea of the World-Soul and refined it down to a soul of the earth itself, and in doing so inspired later thinkers such as Gilbert and Kepler.[27] It is important to note that Ficino is talking about a soul of the element earth, which is differentiated from a separate soul of elemental water, both of which produce imperfect creatures in their own environs. This earth-soul, containing the seminal principles within it, can then swap places with the celestial bodies as the informing agent in spontaneous generation, with these higher bodies reverting to general or cooperative causes.

For Ficino, the earth truly is a living creature, with stones for teeth and plants for hair. He also makes an argument akin to that which Averroes had made for the ensoulment of the celestial bodies. 'Who would say that the womb of this mother lacks life', asks Ficino, 'when of her own accord she brings forth and nourishes so many offspring?' (Ficino 2001, I, 249; Hirai 2008, 287). A similar argument about the generative capacity of the earth was made by Cicero, who directly linked the action of the earth in sustaining vegetative life, as well as its own sustenance from the upper elements, to the nourishment given to the heavenly bodies by the earth's exhalations (*De natura deorum*, II.33; Cicero 1896, 61). We see, therefore, Aristotle's imprecise statements about the role of the sun in generation, and the ubiquity of a certain soul-heat, taken in two partly distinct, partly complementary, directions. The first portrayed spontaneous generation as the formative action of an animate celestial realm on suitably arranged terrestrial matter; the second gave the credit to the animate nature of the earth itself. In their own way, both perspectives, as will be shown in this and following chapters, encouraged speculation along pluralist lines.

Another feature that linked astrology to pluralism within early modern natural philosophy was the question of providence. In terms of astrology, we see this simply posited by Averroes in the *Epitome of Metaphysics* mentioned above: 'Therefore a truthful general statement must be adopted, according to which all [heavenly] motions [were so designed as to] exercise providence over what is below them in this world.'[28] This was followed by medieval Latin philosophers such as the author of the *Speculum astronomiae* (written after 1260), who depicted astrology as the link between metaphysics and natural philosophy (Rutkin 2002, 36–37). The value of astrology as a window into the secrets of God's creation and plan was eloquently defended later by the famous astrologer Girolamo Cardano (1501–1576) in his commentary on Ptolemy's *Tetrabiblos*:

> Nothing comes closer to human happiness than knowing and understanding those things which nature has enclosed within her secrets. Nothing is more noble and excellent than understanding and pondering God's supreme works. Of all doctrines, astrology, which

---

[27] Plotinus, *Enneads*, IV.4, 22, 26–7; ibid., VI.7, 11; Proclus, *Commentary on Plato's Timaeus*, IV.4. See Hirai 2008, 274–75.

[28] Averroes, *Epitome*, 168 (77), quoted in Freudenthal 2002, 122, n. 32. Averroes was influenced in this respect by Alexander of Aphrodisias' *De providentia*.

> embraces both of these—the apotheosis of God's creation in the shape of the machinery of the heavens and the mysterious knowledge of future events—has been unanimously accorded first place by the wise.[29]

The ability to predict the future came from this understanding of 'nature's secrets', the chain of cause and effect that linked the highest and most perfect part of God's creation with terrestrial and human affairs.

Perhaps the most striking and famous example of the providential qualities of astrology is presented by Philip Melanchthon (1497–1560), an admirer of Cardano, whose influence over the creation and administration of Lutheran universities resulted in a so-called 'circle' of mathematicians who advanced astronomy and astrology in a pious attempt to approach God through his celestial secondary causes.[30] According to Melanchthon, astrology is the part of physics 'which teaches what effects the light of the stars has on simple and mixed bodies, and what kind of temperament, what changes and what inclinations it induces'.[31] This simple invocation of the role of celestial light in sublunary processes underlies his conviction that behind the perfectly ordered motions of the celestial realm lies God's plan for the governance of the universe.[32]

The teleological dimension of astrology will be investigated further in later chapters. For someone like Melanchthon, the commitment to an Aristotelian/Ptolemaic cosmology, with its set hierarchy of spheres and emphasis on the instrumental function of the planets and stars, discouraged the entertainment of unorthodox ideas, such as Copernicanism and pluralism, that could disrupt this relationship. This is explicit in the passage where he specifically criticises the theory of a plurality of worlds:

> God is a citizen of this world with us, custodian and server of this world, ruling the motion of the heavens, guiding the constellations, making this earth fruitful, and indeed watching over us; we do not contrive to have him in another world, and to watch over other men also....[33]

The teleology of the heavens, according to Melanchthon, is founded solely on a principle of influence, which is incompatible with the idea of other worlds. He had, of course, more obvious theological objections to such innovations, based on the absurdity and blasphemy of the idea that Christ may have lived, died and been resurrected multiple times on multiple worlds. But then there were just as many theologi-

---

[29] Cardano 1554, fol. A2^r. Quoted in Ernst 1991, 252. See also Ernst 2001.

[30] See Westman 1975; Kusukawa 1995; Brosseder 2005. On the role of astrology in the Reformation more generally, see Barnes 2015.

[31] Melanchthon, *De dignitate astrologiae*, in *Corpus Reformatorum*, XI, 263: 'Astrologia pars est Physices, quae docet, quos effectus astrorum lumen in elementis et mixtis corporibus habeat, qualia temperamenta, quas alterationes, quas inclinationes pariat.' Translated in Methuen 1996, 396.

[32] Questions about the nature, extent and theological dimensions of celestial influence were far more complex in the medieval and early modern periods than is being presented here. A good place to start is the chapter on 'Celestial Influence' in Grant 1994, 569–617.

[33] Melanchthon 1550, fol. 43. Translated in Dick 1982, 89.

cal objections to astrology.[34] Overall, however, Melanchthon's approach was in line with the prevailing opinion that the celestial bodies should be understood in terms of their effect on, and service to, a passive, central earth. In the Renaissance and early modern periods, this picture would be disrupted by the development of cosmological alternatives to Aristotle which triggered a reassessment of the astrological chain of causation, and from that followed a reconsideration of the chain of being, and the contemplation of ET life. It is to the history of the 'plurality of words' philosophy that we now return.

2.3 Nicholas of Cusa and Platonic Influences

From Plutarch, our discussion of pluralism leaps over a millennium to Cardinal Nicholas of Cusa (1401–1464). Provided with the appropriate qualifications and exceptions, we may conclude that medieval Christian scholastic philosophy was dominated by Aristotle to an extent that largely precluded pluralism, or, for that matter, any extension to astrological theory which proposed that the effects as well as the causes of celestial influence could exist in the heavens. The view of the cosmos that dominated Western thought in the Middle Ages and, indeed, up until the late seventeenth century, rested upon a singular enclosed world ontologically demarcated between sublunary and celestial.

That is not to say that certain anti-Aristotelian ideas were not entertained, discussed, and even upheld. The most obvious in terms of its relevance to pluralism was the debate about whether God created more than one universe. It was one of the sticking points between Christianity and Aristotelian philosophy, and a subject of controversy between the Church and the universities at the time of Albertus and Thomas Aquinas. Christian theology dictated, on the one hand, that God created only one universe, while on the other hand it was made a matter of orthodoxy that he could have created more (Duhem 1985, 441–54; Funkenstein 1986, 140–50). This scholastic debate on the distinction between God's *potentia absoluta* and *potentia ordinata* is the main focus of the histories of pluralism in the medieval period.[35] There were, however, other interesting ideas being considered. Robert Grosseteste, for example, posed this question in the thirteenth century:

> But how is it known that there are not more planets, invisible to us but nevertheless useful and necessary for generation in the lower world? For the philosophers say that the Milky Way is made up of very small fixed stars, invisible to us. Therefore, how can it be known, except by divine revelation, whether there are not more stars of this sort invisible to us? For

[34] For a look at some of the main theological controversies up to and including Cardano, see Vescovini 2014.

[35] According to Dick, the crucial development in this period was from 'outright rejection of other worlds to the insistence that they were possible according to natural law' (Dick 1982, 23). See also the rest of his chapter on the medieval tradition, in ibid., 23–43, and also Grant 1994, 150–68.

> the stars which make up the Milky Way, although they cannot be distinguished by sight, are not without effect on generation and growth in the lower world.[36]

This passage occurs amidst Grosseteste's discussion of the mechanics of celestial motion, and his scepticism that a motion of the firmament is necessary for the motion of the planetary bodies. Thus, we have instances of divergence from Aristotelian science, in which the author stresses his ignorance (and that of others) in terms of both the number of celestial bodies and the means of their motion. We should notice, however, that the terms in which the argument is expressed are rooted firmly in a teleology of influence—all celestial bodies, however many there may be, have the function of causing change in the sublunary realm.

Ignorance was a theme of central importance to Nicholas of Cusa, and to the cosmological suggestions presented in his famous work *De docta ignorantia* ('On Learned Ignorance', 1440).[37] Just like Grosseteste, Cusa's stress on ignorance refers largely to the unjustified assumptions and conclusions which form the foundation of medieval cosmology. Many things claimed to be 'known' were based upon spurious observation or reasoning, or were a futile attempt to reduce the scope of unintelligible ideas. According to Cusa, the ordering of the heavens was one example of something that could not be known in a precise way.[38] Although it cannot be proven that Cusa read Plutarch's work on lunar theory, he was competent in Greek, and the similarities between some of the cosmological arguments of *De docta ignorantia* and those in *De facie* make it very tempting to suppose that he had studied the work, possibly during his stay in Constantinople in 1437–38.[39] It was after all on the return voyage that Cusa claimed to have experienced the divine illumination which led him to write *De docta ignorantia* (Cusa 1932, 163).

Cusa's philosophy and cosmology, in particular his concept of infinity and the coincidence of opposites, have already been examined by several specialists in the field (Lovejoy 1948, 112–15; Koyré 1957, 6–24; Dick 1982, 40–42; Achtner 2005; Watanabe et al. 2013), so we can limit ourselves to a consideration of those passages that are relevant to the topics of celestial influence and inhabitation. To begin with, we can look at a passage that describes a Platonic rationale for motion which resembles that of Plutarch:

> For while all things are moved individually, so that they may exist as they are in the best way possible, and so that none moves exactly as another, each thing also contracts and participates in, both mediately and immediately, the motion of each other thing—just as the elements and elemental bodies do in regard to the motion of the heavens, and all the members of the body in regard to the motion of the heart—so that the universe may be one. Through this motion things exist in the best way that they can, and they move for this reason: so that they may be conserved in themselves or in species by means of the natural

[36] Robert Grosseteste, *Hexaemeron*, III, vi, 1–3; viii, 3. Translated in Dales 1980, 541.

[37] Citations will be to the edition in Cusa 1932. Available English translations are Cusa 1985, 1997, 85–206. The translations here have been altered from those in Cusa 1985.

[38] Cusa 1932, 61: 'Caeli etiam dispositio … praecise scibilis non est.'

[39] It is known that he did acquire a copy of Plutarch's *Vitae* and *Moralia* on this trip (British Library, Cod. Harl. 5982), although this particular manuscript does not contain *De facie*.

union of different sexes. These sexes are contracted separately in individuals and united in nature by the enfolding of motion.[40]

There are many elements of this passage that need to be unpacked. First of all, Cusa asserted that things are moved individually according to their own unique purpose. This immediately contradicted the Aristotelian theory of movement which assigned to each thing a motion according to place. This purpose that he alludes to is betterment, which we saw above in Plutarch: a tendency towards the 'good'. Yet Cusa qualified this by asserting further that everything is connected—that they contract and participate in the motion of everything else. This notion of 'contraction' is crucial to Cusa's philosophy, although he never gave a precise definition.[41] To explain simply how Cusa used it in a cosmological and cosmogonic sense, we could say that the universe is perfect and unified in the mind of God, but in reality it is imperfect and contracted into individual species.

This communication of motion between individuals is how the disparately contracted universe can still be considered as 'one', and the examples that Cusa gave were how the members of the body are connected to the motion of the heart, and how elemental bodies are connected to the motions of the heavens. So Cusa justified his theory of motion with both a microcosmic and a macrocosmic analogy, and in terms of the macrocosm this motion of contraction and participation was astrological. Here, even in Cusa's unorthodox cosmology, we can plainly see the appeal of astrological thinking: it creates one out of many—a singular ordered and interconnected world out of the incomprehensibly vast and chaotic. Cusa, however, placed this astrological principle within a providential design that was more unique and immediate than the general and external system of Aristotle. That is, each thing is *individually* motivated to conserve itself and exist in the best possible way, rather than all motion proceeding down from the circumferential prime mover.

The remaining statements by Cusa which pertain to this discussion all derived from his attempt to argue that the earth is not the lowliest, nor the most imperfect or ignoble part of the heavens. One of the prevailing opinions he had to counter was that the earth is dark. This, like many other false conceptions (e.g. the earth's centrality and immobility), is only a matter of perspective. For the sun, thought Cusa, seems to possess regions like the earth, i.e. a central earthy part, a middle watery and airy region, and a fiery circumference. If one were far enough away from the earth to be able to perceive the circumference of its own sphere of fire, then the earth too would look like a bright star. That the moon is similarly dull can be explained by the fact that we are closer to it and are looking directly at its watery region (Cusa 1932, 105). Not only does this prevent us from seeing the bright sphere of the moon,

[40] Cusa 1932, 98: 'Nam dum omnia moventur singulariter, ut sint hoc, quod sunt, meliori modo et nullum sicut aliud aequaliter, tamen motum cuiuslibet quodlibet suo modo contrahit et participat mediate aut immediate, sicut motum caeli elementa et elementata et motum cordis omnia membra, ut sit unum universum. Et per hunc motum sunt res meliori quidem modo, quo possunt. Et ad hoc moventur, ut in se aut in specie conserventur per naturalem sexuum diversorum connexionem, qui in natura complicante motum sunt uniti et divisive contracti in individuis.'

[41] See Hopkins' introduction to Cusa 1985, 17.

but it also prevents us from receiving the heat from its fiery circumference. 'The earth', he wrote, 'seems to be situated between the regions of the sun and the moon, and through the medium of these two bodies it partakes of the influence of other stars'.[42] All these stars, by which Cusa seems to have meant all celestial bodies, are of a similar construction, but because we only see the region of each that shines, we cannot perceive the bodies themselves.[43]

The earth, therefore, being a star like the others, emits its own distinct light, heat and influence. These influences are then shared between the stars, creating a system in which each individual body strives towards the best possible existence, but does so 'in communion' with others: a cosmic ecosystem grown out of the survival instincts of individual stars (Cusa 1932, 106). Cusa expanded on this matter, making it clear that we should not think the earth lowly because it receives influences. It also influences the other celestial bodies, e.g. the sun, but we obviously do not experience this counter-influence (*refluentia*). He then supposed a specific system of mutual influence between the earth, moon, and sun, with the earth being potentiality, the sun being actuality, and the moon being the middle link between the two. In this relationship, the one body cannot exist without the others, and in each body the influence is both three and one. Cusa obviously derived this system from his commitment to Trinitarian theology and its manifestation in the universe. Interestingly, it led him to suggest that these three bodies are in the same region, and that Mercury and Venus are above the sun, 'as some ancients and even some moderns said' (Cusa 1932, 107). This conflicts with another passage in *De docta ignorantia*, in which Cusa seemed to maintain the standard Ptolemaic ordering (Cusa 1932, 102). It is this passage which has led most commentators to assume that while Cusa's cosmology was in many ways revolutionary, his planetary order was conventional.[44] However, this alternative ordering with Mercury and Venus above the sun is part of a more explicit and metaphysically grounded argument by Cusa, which suggests that it should be given more weight in our consideration of his astronomical ideas.[45]

There is one more crucial way in which Cusa's earth was no less noble than the other stars, and it is the main reason why Cusa has such a prominent role in the history of pluralism: the nature of its inhabitants. Cusa believed that all the places in the heavens and stars cannot be empty, but rather that 'natures of diverse nobility' proceed from God and inhabit every region (Cusa 1932, 108). Yet, whatever varying grades of nobility they might engender, these natures cannot be more perfect than that intellectual nature which resides in the earth and its surrounds. But what can we know about these extraterrestrial beings? Not a lot, but Cusa did have some conjectures:

---

[42] Cusa 1932, 105: 'Unde ista terra inter regionem solis et lunae videtur situata et per horum medium participat aliarum stellarum influentiam…'

[43] This seems to be what Cusa means by '…quas [stellas] nos non videmus propter hoc, cum extra earum regiones simus; videmus enim tantum regiones earum, quae scintillant' (Cusa 1932, 105–06). The other alternative is that he means other 'dark' stars which do not shine.

[44] See Hopkins' introduction to Cusa 1985, 27.

[45] But again, Cusa would stress our ignorance about astronomical specifics such as these.

> We are able to know disproportionately less about the inhabitants of other regions. Those in the region of the sun are presumably more solar, meaning more bright, illustrious and intellectual, and even more spiritual than those in the moon, where they are more moon-like (*lunatici*), and those in the earth, which are more material and dense. In this case those intellectual solar natures would exist more in actuality and less in potentiality, while the terrestrial natures would be more in potentiality and less in actuality, with the lunar natures fluctuating in between the two. We suppose this from the fiery influence of the sun, the watery and airy influence of the moon, and the weighty and material influence of the earth. We similarly suppose that none of the other regions of the stars lack inhabitants, so that there may be as many world-like parts of the one universe as there are stars, of which there is no number.[46]

In this passage, Cusa assumed that the nature of a celestial body's inhabitants matched that body's influence—not just in terms of fiery or watery, but also in line with the potentiality/actuality spectrum he had outlined previously. By extension, all the stars (which of course includes the planets) have inhabitants which assumedly conform to their nature; a nature which we perceive manifested in their influences.

Cusa's reliance on the role of celestial influence in generation is continually evident in *De docta ignorantia*.[47] On the topic of earthly corruption, for example, he questioned not only whether corruption is a peculiarly terrestrial phenomenon, but also whether anything can be said to be altogether corruptible at all. Generation is, in his view, a process of the contraction of stellar influences into an individual, and so corruption is just the separation of those influences, which is not so much death as the end of a particular way of being (Cusa 1932, 108–09). Nowhere did he make the connection between astrology and pluralism more apparent than in this question:

> Certain people have said that there are as many species of things in the earth as there are stars. If the earth thus contracts the influence of all the stars into individual species, why does it not happen similarly in the regions of other stars, which receive the influences of the others?[48]

Cusa used the universal nature of his Platonised astrological influence to argue that similar processes of contraction and generation occur on the other celestial bodies. When he imagined extraterrestrials it was through an astrological lens. This lens informed not only his conjectures about their natures, but also his reasoning for *why* and *how* they exist at all. When ET life was discussed seriously for the first time in over a 1000 years, it did not conflict with astrology; to a certain extent it was derived from it.

---

[46] Cusa 1932, 108: 'suspicantes in regione solis magis esse solares, claros et illuminatos intellectuales habitatores, spiritualiores etiam quam in luna, ubi magis lunatici, et in terra magis materiales et grossi; ut illi intellectuales naturae solares sint multum in actu et parum in potentia, terrenae vero magis in potentia et parum in actu, lunares in medio fluctuantes. Hoc quidem opinamur ex influentia ignili solis et aquatica simul et aërea lunae et gravedine materiali terrae, consimiliter de aliis stellarum regionibus suspicantes nullam inhabitatoribus carere, quasi tot sint partes particulares mundiales unius universi, quot sunt stellae, quarum non est numerus…'

[47] On Cusa's astrology, see Roth 2001a, b.

[48] Cusa 1932, 109: 'Dixerunt quidam tot esse rerum species in terra, quot sunt stellae. Si igitur terra omnium stellarum influentiam ita ad singulares species contrahit, quare similiter non fit in regionibus aliarum stellarum influentias aliarum recipientium?'

2.4 Powers from an Infinite Heaven: Palingenius and Digges

Cusa is perhaps better known in the history of cosmology for his ideas on the infinite, and more specifically his argument for a universe without a defined limit. The full ramifications of this idea for pluralism and the ET life debate would be spelled out by Giordano Bruno in the late sixteenth century, but in between these two figures are two others who deserve consideration: Marcellus Palingenius (c. 1500–c. 1551) and Thomas Digges (c. 1546–1595). The thoughts of these two men on the size and structure of the universe were very influential, especially in an English context. Palingenius' *Zodiacus vitae* (1536)—a cosmological poem divided into 12 books, one for each sign of the zodiac—was translated into English in 1560 and became a popular textbook in schools.[49] Thomas Digges' *Perfit Description of the Caelestiall Orbes*, included in a new edition of his father's astrological almanac, offered a rendering in English of the basic tenets of the Copernican system, and was printed at least seven times between 1576 and 1605.[50] Together these two works familiarised audiences with the concept of a universe that was not circumscribed by any absolute limit.

These universes, however, were not infinite in the way Bruno's cosmos would be, or indeed like Cusa's was. Palingenius' cosmos was geocentric, and it retained the strict distinction between the sublunary and superlunary worlds. The infinity he described was the unbounded space which lies beyond the fixed stars, where everything is immaterial and made completely out of light—a 'filling in', in a way, of the Stoic extra-cosmic void. As Alexandre Koyré put it, 'it is God's heaven, not God's world, that Palingenius asserts to be infinite' (Koyré 1957, 27). Even when the poet argued that the heavens are filled with God's creatures, he made a sharp distinction between those perfect and immortal beings and the mortal, corporeal creatures on earth. Yet his emphasis on the infinite seems to have been motivated by the same 'principle of plenitude' that Lovejoy identified as the main driving force towards pluralism. Whatever God could do, we must believe he actually *did*, lest his power remained idle, vain, and unused (Palingenius 1996, 477).

The universe thus described, as the name of the poem would suggest, is highly astrological. 'What virtues have the heavens?', asked the poet. 'All power is in the stars: they rule the orb of the earth and change everything. They create and govern all things in the earth.'[51] Palingenius' astrology, like Cusa's, was Platonised. The causes and 'seeds' of all things lie in that highest, immaterial heaven. In that place is the mental archetype from which flows the sensible world as an 'image' (Palingenius 1996, 255). He defended the 'divine mind' of Plato, who knew of these incorporeal forms, against the venomous mob which slandered and mocked them

[49] For the English translation, see Palingenius 1947 [1560]. References here will be to the Latin text in Palingenius 1996.

[50] See Johnson and Larkey 1934.

[51] Palingenius 1996, 443: 'Quid virtutis habet coelum? Est vis omnis in astris: Astra regunt orbem terrarum, atque omnia mutant: Astra creant cuncta in terris, et cuncta gubernant.'

(Palingenius 1996, 481). While his poetry emulated Lucretius' *De rerum natura*, which had been rediscovered in the fifteenth century, he was scornful of the 'sophistic' atomist philosophy. He offered a Platonic compromise: perhaps we shouldn't think of atoms as bodies, but rather souls, which could likewise be called a celestial seed, and an eternal progeny of God (Palingenius 1996, 271).[52] His theory of animal motion was similarly Platonic. The mind sees 'good', and then moves towards it, but at the same time is drawn by the virtue of the object, like fire to fuel, iron to the magnet, or chaff to amber. The dual notions of friendship and strife are thus the origin of all action. They bring the will into motion, which in turn drives the limbs and impels the body (Palingenius 1996, 259).

Palingenius argued that the heavens are inhabited—that, considering the plethora of life on earth, it is madness to think that all those vast spaces are empty (Palingenius 1996, 365). These celestial beings are not subject to generation and corruption, as was mentioned earlier, but, interestingly, the author did draw an analogy between them and terrestrial creatures. He did this in anticipation of the criticism that no creatures could move or subsist in the solid diamond spheres of the heavens:

> Who, unless he saw it, would believe fish to live under the waves, frogs to live under mud, salamanders to live in fire, chameleons to be nourished by air and cicadas by dew? Yet we confess these things to be true and wonderful. There are many things, which although we believe them to be impossible, often can be done and are seen to be so. Therefore why could God not also create Heaven-dwellers of such a kind that are able to pass easily through the heavens, needing no food or drink at all?[53]

Palingenius was creating a comparison here between celestial life and unusual terrestrial biology, arguing, similarly to Plutarch, that the variety found in nature demonstrates God's ability to fit suitable inhabitants to every region, on earth and beyond. So, while the *Zodiacus vitae* does not, as Cusa's *De docta ignorantia* had done, consider the earth and the planets to be varieties of the same kind, it did present a vision of the celestial cosmos with the capacity for both influence and inhabitation. Like his treatment of the infinite, the author's portrayal of heavenly beings came close to being an assertion of ET life, but, in the end, it was not. This is not 'life' as we would think of it. Pluralism was treated by Palingenius much as he treated other ancient philosophies, reinterpreting them according to his own Christian theology while downplaying their material reality.[54]

Another passage in the poem which lends it a progressive feel is where Palingenius famously stated: 'Some believe it is possible to call every star a world, and they

---

[52] On Palingenius and Lucretius, see Haskell 2015. See also Bacchelli 1999, 2001.

[53] Palingenius 1996, 451: 'Quis, nisi vidisset pisces habitare sub undis, sub limo ranas, salamandras vivere in igne, aëre chameleonta, et pasci rore cicadas, crederet? At vera haec tamen et mira esse fatemur. Plurima sunt, quae cum fieri non posse putemus, saepe tamen fieri possunt, et facta videmus. Cur non ergo Deus potuit quoque condere tales, Coelicolas, qui per coelum facile ire valerent, nulliusque cibi vel potus prorsus egerent?'

[54] This was argued by Foster Watson in 1908, who was relying on earlier work by Gustave Reynier on the sources of the *Zodiacus vitae*. Reynier had found that much of the occult material in the poem was taken from Agrippa's *De occulta philosophia* (1529). Watson saw Palingenius' spiritual approach to 'older philosophies' as similar to Agrippa's (Watson 1908, 69–70).

declare the earth a dark star.'[55] This line, and the poem as a whole, seems to have had a great influence on the Englishman Thomas Digges, who quoted it in his *A Perfit Description of the Caelestiall Orbes*.[56] When this line is quoted in the secondary literature, its context is rarely discussed. It came in the middle of a passage which was affirming the distinction between the celestial and the terrestrial. The poet compared the stars to carpenters, framing matter into diverse forms; forms which can never be more than shadowy reflections of their perfect counterparts. Digges, in his account of the Copernican system, is credited as the first to make the so-called logical step of breaking the solid orb of the fixed stars and extending it infinitely upwards. But even while making the earth orbit the sun with the other planets, he distinguished it as the specific orb of mortality, made up of the four elements, citing Palingenius in support (Johnson and Larkey 1934, 80–81).[57] This 'darkened starre' remains unique, and the infinite region of the stars is described as the eternal court of celestial angels, and is thus ontologically distinct from the world below.

Palingenius and Digges, while retaining important features of the terrestrial/celestial divide, spread the concepts of an infinite universe and celestial inhabitants to a wide audience. What often gets overlooked or dismissed in the history of these figures, however, is that these innovations were presented in an astrological setting. 'The Zodiac of Life', divided into 12 books bearing the names of the signs of the zodiac, is above all an ethical work. It talks at length about celestial inhabitants (*coelicolae*) while continually reminding the reader that in that same superior world lie the causes of all things here below. Digges' treatise, as mentioned above, was included in his corrected edition of his father's *A Prognostication Everlasting* (first published in 1555), a perpetual astrological almanac which contained 'chosen rules to iudge the weather by the sunne, moone, starres, comets, rainebow, thunder, cloudes, with other extraordinary tokens, not omitting the aspects of planets, with a briefe iudgement for euer, of plenty, lacke, sickenes, dearth, warres &c. opening also many naturall causes worthy to be knowen' (Leonard Digges 1576, 1). Astrology dealt with the nature of the celestial bodies, and so naturally it shared ground with pluralist ideas such as inhabitation and infinity. While the main purpose of the heavens remained that of celestial influence, there was sufficient scope within this paradigm to allow for the introduction of ideas from the rediscovered Platonic, Stoic and atomistic philosophies, or the new Copernican astronomy. In the expanded universes of Palingenius and Digges, celestial inhabitation is given more importance, but that kind of life that involves generation and corruption remains confined to the earth.

---

[55] Palingenius 1996, 255: 'Singula nonnulli credunt quoque sidera posse dici orbes, terramquae appellant sidus opacum…'

[56] *A Perfit Description* is reproduced in Johnson and Larkey 1934, 78–95. On Digges and Palingenius, see ibid., 101–4.

[57] See also Roos 2001, 88.

2.5 Pluralism and Italian Anti-Aristotelianism: Giambattista Benedetti and Francesco Patrizi

The cosmological innovations of Nicholas Copernicus (1473–1543) and Tycho Brahe (1546–1601) are central to the history of the ET life debate. Copernicus' heliocentric theory made the earth one of the planets, opening up the possibility of analogies between terrestrial and planetary natures. The earth's displacement from the centre of the universe also necessitated a rethink of theories about natural motion and gravity.[58] Tycho's geo-heliocentric alternative—with the planets orbiting the sun which in turn orbits the central earth—didn't require such a drastic upheaval, but it did suggest that the celestial ether was liquid or permeable rather than being composed of solid crystalline orbs. His observations of the new star of 1572 and the comet of 1577 also seemed to prove that at least some level of alteration took place in the superlunary region (Blair 1990; Mosley 2007). As well as obvious direct implications for the question of ET life, the possible motion(s) of the earth and corruptibility of the heavens also had an indirect impact through new developments in theories about celestial influence.

The scope for modification within the celestial influence paradigm is demonstrated in different ways by two Italian contemporaries and correspondents: Giambattista Benedetti (1530–1590) and Francesco Patrizi (1529–1597). These authors also provide evidence for the way in which astrological questions could encourage pluralist ideas. Benedetti was a Venetian mathematician best known for his theories of impetus which later influenced Galileo. He served at various times as the Court Mathematician to Duke Farnese of Parma and later to the Dukes of Savoy in Turin. These posts included duties as an astrologer, and indeed Benedetti maintained a lifelong interest and faith in judicial astrology. He had predicted his own death for the year 1592, but on his death bed in 1590 he recalculated and concluded that the original nativity must have involved an error of 4 min (Drake 1970). He was also involved in an astrological dispute with one Benedetto Altavilla of Vicenza, in which he defended the reliability of ephemerides and the validity of prognostication.[59]

Benedetti argued for the superiority of the Copernican ephemerides for use in astrology, but rather than taking a 'Wittenberg interpretation' of Copernicus, he was also enthusiastic about the reality of the system itself.[60] His cosmological views can be found scattered throughout his later work, *Diversarum speculationum mathematicarum et physicarum liber* ('A Book of Various Meditations in Matters of Mathematics and Physics'), published in Turin in 1585, where, among other anti-Aristotelian views, he rejected the system of solid spheres in favour of planets moving independently through the celestial air according to providentially determined

[58] On Copernicus' impact on pluralism, see Dick 1982, 66–105.

[59] See 'Defensio Ephemeridum', in Benedetti 1585, 228–48. See also Omodeo 2014, 145–49.

[60] See Westman 1975. Westman characterised this interpretation as involving the adoption of Copernican mathematical models and ephemerides, while ignoring the physical implications.

paths (Omodeo 2014, 30–31). He also wrote an essay entitled 'The opinion of those who believe there to be many worlds is less than sufficiently rejected by Aristotle', in which he countered specifically the argument that the elements of another world would seek their natural place in this one.[61] If these are indeed 'worlds' in the proper sense of the term, they will have their own centre and circumference.

This led Benedetti to then invoke the opinion of Aristarchus, or at least his own unusual interpretation of it:

> For example, if the opinion of the most learned Aristarchus is true, it would stand to reason that what happens in the case of the moon may happen similarly to each of the other five planets. That is, as the moon revolves around the earth by means of its epicycles, as though [moving] through the circumference of a second epicycle, in which the earth takes the form of a natural centre (i.e., it is in the middle), carried by the annual orbit around the sun; in the same way also Saturn, Jupiter, Mars, Venus and Mercury may revolve around some other body, which is situated in the middle of their own major epicycle. This body may also have a certain motion around its own axis, be opaque, and be furnished with conditions similar to the earth; while in the epicycle, as we have called it, things may be similar to the way they are on the moon.[62]

As we can see, Benedetti believed that in the 'Samian' or Copernican system the planets should be thought of as moons, orbiting around other earths which are 'opaque', meaning dark, and therefore invisible to us.[63] This is an example of a Renaissance astrologer who took a keen interest in both physics, metaphysics, and cosmology. It seems that Benedetti was trying to give a physical account or reason for the main planetary epicycles that remain in the Copernican system. The moon, which orbits in epicycles of its own around a point (the earth) on a deferent circle (the earth's orbit around the sun), seemed to offer the perfect model.

Benedetti discussed this idea again in his *Diversae speculationes* in the form of a published letter to the Italian historian Filiberto Pingone (1525–1582), entitled 'On the purpose of the celestial bodies and their motion.' This short letter sheds light on Benedetti's model of celestial influence and its relation to his peculiar 'Copernican' pluralism. It begins:

> If you desire to know to what end, apart from superior light, the celestial bodies have been made, and if you wish to follow human reason, considering that these bodies are divine, or of such an incomprehensible number, with magnitudes so great and motions so fast, then you should not believe that they have been made only to rule such a vile body, as is the earth encompassed by water, with its animals and plants. Those who follow the opinion of

[61] Benedetti 1585, 195: 'Minus sufficienter explosam fuisse ab Aristotele opinionem credentium plures mundos existere.'

[62] Benedetti 1585, 195–96: 'ut exempli gratia, si doctissimi Aristarchi opinio est vera, rationi quoque consentaneum erit maxime, ut quod lunae contingit, cuilibet etiam ex aliis quinque planetis eveniat, idest, ut quemadmodum Luna suorum epicyclorum ope circum terram voluitur, quasi per circumferentiam alterius cuiusdam epicycli, in quo terra sit instar centri naturalis (idest sit in medio) delati ab orbe annuo circa Solem; Sic etiam Saturnus, Iupiter, Mars, Venus, atque Mercurius, circum aliquod corpus in medio sui epicicli maioris, situm habens, volvantur; quod quidem corpus, et aliquem quoque habeat motum circa suum axem, sit opacum, iis conditionibus, quae terrae sunt similes, praeditum existat, et in dicto epyciclo sint res similes istis lunaribus.'

[63] On Benedetti's interpretation of Copernicus, see also Omodeo 2014, 175–78.

> Aristarchus of Samos and Nicholas Copernicus will believe this even less. According to them it cannot be the case that the rest of the universe has no other purpose than the rule, to use their words, of this centre of the epicycle of the moon. If this opinion is true (as they reasonably consider it to be), how unseemly would it be if the centres of the epicycles of the other planets were deprived of similar governance; indeed, it consents in no way with reason.[64]

Elaborating on this teleological description of the celestial bodies, Benedetti goes on to repeat his cosmological theory: that the planets are moons orbiting around unseen 'earths' (Benedetti 1585, 255). It is interesting to see a practitioner and defender of astrology state so strongly that the governance of the earth cannot be the only purpose for the creation of the planets and stars. The greatly increased size of the universe in the Copernican system required to account for the lack of stellar parallax encouraged sympathetic readers to consider philosophical alternatives to Aristotle, but, as we can see, Benedetti did not abandon the concept of celestial influence, rather he expanded it along with the cosmos.

Doing so required him to counter the argument of Aristotle that the celestial regions do not experience generation and corruption, and have never been seen to change—a feat he accomplished in his letter by explaining that earthly change only occurs in small magnitudes, and from a distance its pattern of earth and sea would look as stable as any other celestial body (Benedetti 1585, 255). Much of the rest of the letter is an argument for Copernicanism, claiming that the axial rotation of the earth is preferable to the excessive speeds required by the diurnal rotation of the planets, and that an annual orbit of the earth easily substitutes for that of the sun. The participation of the earth's parts in its natural impetus counteracts Ptolemaic objections, and this theory also, in Benedetti's mind, removes the need for a finite universe. It is interesting that he saw the need to argue that both the diurnal and annual motion of the earth suffice to expose it to the light, heat, and influence of the sun and the other bodies (Benedetti 1585, 255–56). The celestial influence received by the earth would not change whether you assumed that it moved or not, yet we see this argument repeated again and again by early Copernicans.

Furthermore, the provision of celestial influence was the key factor in Benedetti's repetition of his 'lunar planet' theory. His theory of other earths in the centre of the planetary epicycles was the answer to Pingone's question about the purpose of the celestial bodies, and of their motion. As he explained in another letter, this time to Antoine of Navarre (1518–1562), the rationale behind the motion of the celestial

[64] 'De fine corporum coelestium, et eorum motu', in Benedetti 1585, 255–56: '... si absque lumine superiori, in quem finem facta fuerint corpora coelestia scire desideras, et humanam rationem sequi volueris, putandum tibi non erit ea. solum effecta esse, ut tam vile corpus, ut est terra aquis irrigata, animalia, et plantas regant, cum ea. corpora sint divina, an numero incomprehensibilia, maximis magnitudinibus, et motibus velocissimis, praedita, id etiam minus putabunt hii, qui opinionem Aristachi Samii, et Nicolai Copernici sequuntur, quorum ratione fieri non potest, ut credant; eius, quod ex uni[verso reliquum] est, alium finem non habere, quam regimen huius centri epicycli Lunaris, ut illorum more loquar. Quam enim turpe esset si centra aliorum epicyclorum planetarum tali regimine privarentur, id quod nullo modo cum ratione consentit, [si tam] vera est ea. opinio, quemadmodum rationabiliorem eam existimant.'

bodies is the need to diversify influences.[65] His system reclassified the planets as moons, all of which act, as he explained to Pingone, 'like mirrors conferring the light of the sun to their centre by reflection'.[66] This repetition of the earth/moon partnership explained for Benedetti not only the epicyclic motion of the planets, but, by extension, the ends to which they were created: as agents of celestial influence in a multi-earth cosmos. As mentioned above, it should not be surprising that discussions by astrologers, or about astrology, ventured into the realms of celestial physics, metaphysics, and indeed pluralism. By the same token, it is not surprising that these discussions of pluralism were influenced by the astrological paradigm.

We see similar trends in the innovative cosmology of another Venetian, Francesco Patrizi, who was a correspondent of Benedetti. Patrizi was a defender of Plato against Aristotle who held the chair of philosophy at the University of Ferrara, and then later the chair of Platonic philosophy in Rome. Some of his most influential theories concerned optics and light metaphysics.[67] Being influenced in turn by the Platonic theories of Plotinus, Proclus and Ficino, Patrizi believed that light was the preeminent cosmic force, and drew a distinction between *lux* (light) and *lumen* (the brightness which emanates from it).[68] Light was for him a formal, efficient, and material cause, the basis of both existence and behaviour for created things (Brickman 1941, 28; Kristeller 1964, 119). His work, true to the pattern of Renaissance Platonism, demonstrates an interest in magic and occult philosophy, and he published translations of Zoroastrian and Hermetic texts under the title *Magia philosophica* (Hamburg, 1593).[69]

The work that interests us is his *Nova de universis philosophia* ('A New Universal Philosophy', 1591). The work is divided into four books, with the longest, and most relevant to this discussion, being the fourth, with the title 'Pancosmia'. In this book, he argued favourably for the rotation of the earth (but not its annual motion) and discussed the possibility of an infinite universe and an inhabited moon. On the latter point, in agreement with the Pythagoreans, he thought it possible that there were mountains and valleys on the moon, as well as giant races of men and animals, and plants more beautiful than here on earth (Patrizi 1591, fols 112^{v}–113^{r}). Yet Patrizi's description of an earth-like moon needs to be qualified. He suggested that perhaps the moon is compounded from the crass sediment of the rest of the ethereal region,

---

[65] Benedetti 1585, 413: 'Motus corporum coelestium fit ratione situs, et varietatis virtutis stellae in diversis locis, haec autem varietas absque diverso situ eiusdem stellae, nec diversus hic situs absque motu fieri posset, ita ut motus stellarum sit ratione diversitatis situum ipsarum, ergo motus, et diversitas situum, fit, ob diversam influentiam.' Translation: 'The motion of the celestial bodies takes place in order that the position and influence of a star vary in diverse parts [of the earth]. This variety, however, cannot be achieved without the diverse position of that star, nor can this diverse position be achieved without motion. Thus, as the motion of the stars occurs on account of the diversity of their positions, the motion and the diversity of positions occurs in order to create a diverse influence.'

[66] Benedetti 1585, 255: 'quasi specula, lumen Solis suo centro ex reflexione, deferentia'.

[67] See Lindberg 1986, 28–29. On issues related to atomism, see Henry 2001.

[68] This was actually a common scholastic distinction. See Grant 1994, 392–93.

[69] See Muccillo 1986; Vasoli 1989.

as the earth is from that of the material world. The earth and the moon therefore make a unique pair:

> Let the moon, therefore, be the ethereal earth. And let our earth be the elemental moon. These two earths, or indeed these two moons, were not made by God in vain. Indeed, according to a mutual similarity, the one supports the other, and they are favourable to each other in turn. They receive influences one from the other, by which they live and are preserved in themselves and in their parts.[70]

Rather than the moon being a ready example to extend to the other planets by analogy, the moon is a unique case: the partner of the earth, as described by the ancient philosophies. This uniqueness of the moon is an issue which we see coming up again and again, and it presents an obstacle to the historian who wishes to extrapolate from lunar theories to broader pluralist ideas.

The problem was summed up by Lambert Daneau (c. 1535–c. 1590) in his popular *Physica christiana* ('A Christian Natural Philosophy', 1576), translated into English as *The Wonderfull Workmanship of the World* (1578). Some people, he tells us, believe that the ethereal region spans from the highest heaven to the moon, under which lies the elementary region. Others, however, believe that the moon should be considered part of the elementary and earthly region: 'concerning which varietie of mens opinions, reade *Plutarches* book of the face whiche appeareth in the globe of the Moone' (Daneau 1578, fol. 58^{v}). In this case, Daneau adhered to the former opinion. For Benedetti, likewise, the moon had not been earth-like, and the marks on its surface were simply areas which were more diaphanous and less reflective (Benedetti 1585, 299). While he postulated the existence of other earths, the moon retained its more traditional nature and role as a luminary. In Patrizi's new philosophy, the moon shares in some of the earth's deficiencies, but this is compensated by the beneficial effects they have on each other:

> And since feculent bodies are by their own nature the most torpid and ignoble, there was a need that, besides the benefits which they draw from the sun (which are specific to each), the moon and the earth should aid each other mutually. This they do by sending their own powers one to the other. That being the case, while it is generally known to us what the moon does in the earth, it is not known what the earth does in the moon.[71]

In the final analysis, then, the existence of certain beings on the moon is affirmed, but it relied, as was the case with Plutarch and Cusa, on the lunar influence on the earth being reciprocated by an earthly influence on the moon.[72]

---

[70] Patrizi 1591, fol. 113^{r}: 'Luna ergo, aetherea terra esto. Et terra nostra, elementalis esto luna. Neque duae hae terrae, sive duae hae lunae, a Conditore frustra sunt conditae. Similitudine enim mutua, altera alteram fovet, et sibi invicem favent, et influxus, quibus et ipsae, et utriusque partes, tum vivant, tum conserventur, altera ab altera suscipiunt.' In this and other cosmological areas, Patrizi was greatly influenced by Proclus. See Fabbri 2012, 2016.

[71] Patrizi 1591, fol. 113^{v}: 'Et quia foeculenta corpora, torpidissima, sui sunt natura, et ignobilissima, necesse habuerunt, ut praeter ea. quae a sole, beneficia hauriunt, sese propriis, luna et terra mutuo iuvarent, altera alteri vires suas influendo. Qua in re, quid terra in lunam agat, sicut est nobis ignotum, sic est fere nobis cognitum, quae luna agat in terram.'

[72] See Fabbri 2016.

Patrizi was not, however, as big a fan of judicial astrology as Benedetti. Following his chapter on the moon, he dedicated a chapter to answering the question 'Are the stars causes?', and stated that most things said about the stars in regard to effects and events were baseless.[73] Yet he did allow that perhaps events may be signified by them, because all the parts of the world are interconnected (Patrizi 1591, fols 115^{r-v}). This position was based on Plotinus' consideration of the same question in *Ennead* II.3, which had been translated by Ficino.[74] Patrizi considered that the attacks on the false astrological arts by Ficino and Pico both relied on this work by Plotinus. And yet while he assented to this criticism of astrological practice, he did not deny that the stars do perform some actions. To understand Patrizi's theories of celestial influence, it is necessary to look a bit closer at his novel natural philosophy.

His cosmology retained a hierarchical division, with the empyrean heaven at the top, followed by the ethereal regions, and finally the 'hylaeal' sublunary regions of air, water and earth. In this scheme, the universe is infinite only on the far side of the empyrean, which encloses the finite material world (Patrizi 1591, fols 80^{r}–82^{r}).[75] This infinite universe is pure light, which descends through the immobile empyrean, becoming heat, and then fire, and forming the seeds of all things below. Below the empyrean heaven Patrizi united, to a certain degree, the terrestrial and celestial worlds. There are four essential ingredients to the entire universe: space, light, heat, and *fluor*. These finite inner regions of the ether and the material world are both mobile and indeed both mutable (Vasoli 2002). In fact, they are both corporeal, being formed out of the one primeval liquid (*fluor*) (Patrizi 1591, fol. 121^{v}). The division of the world into separate regions is based solely on a scale of rarity and density, a scale which descends from the empyrean heaven down to the earth, with rareness being the closest to Godliness (Patrizi 1591, fol. 122^{r}; Rosen 1984).

The stars and planets are individual bodies composed of condensed and incandescent *fluor* within the ethereal regions, just as the air, water and earth are even further condensed sediments in the *hylaeum*. The ethereal and terrestrial regions, therefore, are both complex and corruptible. Yet some sort of division between the ethereal and sublunary regions does remain, perhaps because the difference in density is such that it prevents material interaction. While earth, water, and air freely intermingle, the ethereal region sits on the sublunary world like oil on water. This makes the comparison between the earth and the moon easier to understand. The moon is to the heavens what the earth is to the regions of water and air, a body formed by sediment accumulating and compacting in the lowest region.

At the same time, it demonstrates the limits of any possible pluralism within Patrizi's cosmology. In his chapter on the planets, he argued that the sun should not be counted as one of these 'wandering stars', and that the moon should not be counted as a star at all (Patrizi 1591, fol. 105^{v}). Its terrestrial components, or at least its ethereal analogues, seem particular to it as a result of its place in the cosmos, as

[73] Patrizi 1591, fols 114^{v}–117^{r}: 'An stellae aliquid agant.'

[74] Plotinus, *Ennead*, II.3, 'Utrum stellae aliquid agant', in Plotinus 1580, 137–59.

[75] On Patrizi's theory of space, see Henry 1979; Grant 1981, 199–206.

opposed to the stars and planets which are purer, luminescent bodies. For Patrizi the Platonist, these bodies were of course alive:

> They are not, as the common crowd of astronomers and philosophers believe, inanimate bodies. Rather they are, as Zoroaster rightly called them, and Plato and Aristotle after him, animals. Therefore, by nature, which executes their actions, and by spirit, which is their conveyor, and by soul, which gives them the beginnings of motion, and by intellect, from which comes and depends the order of all things, they themselves live, and are carried, and act, and comply to the will of God, and moderate the harmony of the universe. As is useful for the world, they traverse courses, and execute their other actions, and the seeds and spirit of the varied generation of our things flow in.[76]

This cosmology of ensouled celestial bodies within a hierarchical yet somewhat homogeneous universe forms the setting for Patrizi's chapter on the effects of the stars, which constitutes something of a dividing point in the 'Pancosmia', separating the chapters on the ethereal subjects and those on terrestrial ones. It is immediately obvious that, for Patrizi, the key issue was generation. The introduction to the chapter is primarily concerned with Aristotle's comments on the subject. He quoted Aristotle on man and the sun begetting man, and on the biological spirit analogous to the element of the stars, and criticised him on both counts for his opacity and lack of explanation (Patrizi 1591, fol. 114^{v}). The Philosopher limited his explanation of generation to those things which have seeds, namely animals, and so leaves aside all those zoophytes, plants, and metallic stones which are generated without them. Just as we saw in the medieval period, it is spontaneous generation that again raised the question of celestial influence.

This is why his assent to the criticism of astrologers did not translate to a blanket denial of astrology's major premise. His agreement with Plotinus, Ficino and Pico only extends so far. As he wrote:

> The question is proposed by us whether the stars do anything. *This is something more universal than can be reduced to man alone*, as these three great men did. One could ask, for example, whether the stars administer the care of man alone, or whether they have some good or bad influence only on themselves. But the question must spread more widely: do the stars do anything at all? This action either remains amongst them, or is extended through the entire heaven, or diffuses itself either to all parts of the heaven, singular, near, far, and as far as possible. Either it extends into the superior empyrean world, or descends into the inferior material world, and pervades through the universal air, water and earth, occupying all parts of them. And indeed, whether they cause something in metals, in stones of all types, in plants and zoophytes, in all species of animals, not less than whether they act only in men, should be considered, if not by astrologers, who ply their trade for the catching of riches, then at least by philosophers, whose proper function is to search for truth and inquire into the causes of things.

---

[76] Patrizi 1591, fols 105^{v}–106^{r}: 'Suntque non uti Astronomorum, et philosophorum plerumque vulgus existimavit, corpora inanima. Sed sunt, ut vere eos appellavit Zoroaster, et post eum Plato, et Aristoteles, animalia. Igitur, et natura quae actiones eorum peragit, et spiritu, qui eorum est vector, et animo, qui motus eis dat initia, et intellectu, a quo omnis rerum venit ac pendet ordo, ipsa vivunt, et feruntur, et agunt, et Conditoris nutibus obtemperant, et universi harmoniam contemperant, et ut mundo expedit, cursus obeunt, et actiones alias suas peragunt, et generationis tam variae rerum nostratium, semina, atque spiritus influunt.'

> Let us, therefore, put the matter in the following terms: The stars, by the admission of all astrologers and philosophers, are bodies. Yet all bodies are substances, and all substances have their own powers. Powers come forth in actions proper to themselves. We have previously contemplated the substance of the stars. Now the powers and actions of them need to be discovered. The stars were created by God. God creates nothing in vain. Therefore, the stars being bodies so remarkable, so enormous, and of such a number, who could say he made them in vain? Nor, therefore, did he give them powers in vain, nor from these powers do actions vainly proceed.[77]

Hence it is affirmed that the stars are causes. Yet, crucially, discussion of their effects expands beyond any influence they may have on human affairs to more universal considerations. The philosopher, who can remain unbiased in his judgment, should enquire into their possible effects on all things animal, vegetable and mineral, as well as into the effects they have on one another.

The method of their influence, however, remained to be decided. Aristotle's assertion that this comes from motion and light could not be maintained, because Patrizi had robbed the ether of motion by granting a diurnal rotation to the earth. The celestial bodies do move themselves, but this motion does not reach the earth (Patrizi 1591, fol. 114^{v}). The remaining option then was light. His own theory of their effects was Platonic, based on the idea that the forms, which are the seeds of all things, proceed from the world of archetypal ideas (which he identifies with the empyrean heaven) (Patrizi 1591, fols 115^{v}–116^{r}). These seeds are joined to the light, which serves as a vehicle. Sticking to his quadripartite theory, these seeds are received and contained in space, and diffused by the light into heat and *fluor*, by which all bodies and forms of bodies are affected (Patrizi 1591, fol. 115^{v}).

Before entering the ethereal and hylaeal worlds, however, these seeds act first and foremost within the empyrean heaven itself, producing 'empyrean animals, zoophytes, plants, rocks, metals, and other types of things proper to that realm'.[78]

---

[77] Patrizi 1591, fol. 115^{v}: 'Quaestio est nobis proposita, utrum stellae aliquid agant. Haec quidem universalior est, quam ut ad homines solos, uti hi tres magni fecere viri, redigatur. An scilicet stellae, hominum tantum, vel curam gerant, vel ipsis solis boni, vel mali influant aliquid. Sed in amplius, est quaestio fundenda, utrum omnino agant aliquid stellae? Actio haec, vel inter eas versatur, vel per totum extenditur coelum, vel per partes eius omnes, et singulas vagatur, et proximas, et longinquas, et procul maxime positas. Vel etiam in superum mundum Empyreum porrigitur, vel in inferiorem hylaeum descendit, et per aerem pervadit universum, et aquam, et terram occupat universas, tum horum partes. Nec non etiam, an quid in metalla, in tot generum lapides, in plantas in Zoophyta, in tot animalium species, non minus quam, an in homines tantum agerent, considerare si non Astrologi, qui pecuniarum suam artem fecerunt aucupium, philosophi saltem debuerunt, quibus, veritatis vestigandae proprium est munus, et rerum causas inquirendi. Nos igitur ita dicamus. Astra confessione tum Astrologorum, tum philosophorum omnium, corpora sunt. Corpora aut omnia, sunt substantiae. Substantiae aut omnes vires habent suas. Vires in actiones sibi proprias exeunt. Siderum substantiam iam antea sumus contemplati. Vires modo eorum, et actiones sunt cognoscendæ. Astra a Deo sunt creata. Deus nihil frustra facit. Igitur, astra, corpora tam insignia, tam ingentia, tanto numero, quis dixerit eum fecisse frustra? Neque igitur vires eis indidit frustra, neque ex iis viribus, frustra exeunt actiones.' Emphasis added.

[78] Patrizi 1591, fol. 116^{r}: 'At quid producunt? Empyreos nimirum partus, Empyrea animalia, Zoophyta, plantas, lapides, metalla, reliquorum entium genera, Empyreo convenientia, Empyreo propria.'

From there they descend into the ethereal world and are disseminated through the entire space, preventing the existence of a vacuum by establishing a seminal plenum. And as they are spread through the whole heaven, they also exist in the stars. As these stars are one step down from the perfect 'incorporeal', then so too will the seeds undergo a similar degeneration. But is there one seed in each star, the same seed in every one, or all seeds in all of them? Patrizi's preference was that individual seeds are in individual stars in such a way that all the seeds are spread amongst all the stars.[79] Yet while a star has one seed in essence and species, its actions can be many as the effects are divided into individuals of the same species.

It is confirmed therefore that the stars are not idle, and that they are in fact causes, and moreover the underlying method of causation has been described. But *what* do they cause? 'Actions, which correspond to each of them, and which are unique to each', thought Patrizi, 'partly known to us, and partly unknown'.[80] The known actions are light and heat. As for the less known, Patrizi returned to the question of generation. If it is true that the sun and man generate man, it must also be true that the sun generates individuals of all the other species. And if the sun does this, why shouldn't the other planets, being made of the same substance, do the same? It is here that Patrizi's discussion took an unconventional turn:

> Each of the stars has seeds. Either they project them within themselves, or mutually amongst themselves. Indeed, either they project them, or they don't. This projection, we understand as fertilisation. If they do not fertilise, both the seeds and the stars are superfluous, and have been made in vain by the Maker. But to say that is impious, and so they must fertilise. But do they do so within themselves, or outside of themselves? Indeed it is both. Within themselves in so far as they are fertilised and proliferate in a way that is suitable to them. Our senses, however, do not comprehend that, but reason does. Reason declares, for instance, that God did not make the stars in vain. They would be vain, however, if they did not have any effect. Every one of their seeds, therefore, fertilises first of all within themselves.
>
> And thus we should approve of those Pythagoreans, who affirmed every single star to be a world; worlds which contain their own airs, waters, and earths, and have animals living in them more pure and divine than ours. For that is much more probable than to suggest or assert that all those huge and beautiful bodies were crafted by the Maker in vain, or effectively in vain. Or if no philosophy admits this, it is at least necessary to admit something similar, for instance, that the stars themselves are celestial worlds living by their own spirit, soul and mind, or that they are worlds living entirely by the universal life, from the universal mind, soul and spirit.
>
> Outside of themselves, however, they should act at a distance as far off as they can cast their light, and with this light also the powers of the seeds, by which they aid and favour each other mutually. They should conspire amongst themselves in turn; they inspire good things both to themselves and to the whole; and they should imbue everything with the goodness given to them by the Maker. And they conspire so as to become one; so that all sound aloud and together that harmony of the Divine Artificer. This is not that harmony which Aristotle foolishly laughed at in the Pythagoreans, but that which the Pythagoreans assert in the heavens, and we, it must be said, assert in the whole universe.

---

[79] Patrizi 1591, fol. 116^r: '...ita in stellis singulis semina erunt singula, ita ut in omnibus stellis, semina sint omnia'.

[80] Patrizi 1591, fol. 116^r: 'Actiones, quæ singulis illis competunt, quæque propria sunt cuique, notas partim, partim nobis ignotas.'

> For if the stars were said to act wholly within themselves, one could not say that they have been made in vain; but they would be somewhat in vain if they did not effect something between each other, operate mutually, or assist one another. Or if such great lights poured out in such great number into the material world, yet barely reach it. Or if the stars were believed to be made solely for the benefit of the earth.[81]

The key doctrine that informed Patrizi's discussion of stellar activity was that God creates nothing in vain. As we can see in the last paragraph, the benefits that the stars may give to the earth are not enough to satisfy this condition. This necessitates that the seminal and fertilising powers of the stars must have an effect within the stars themselves, and indeed this would be enough to justify their creation. The preferred option, though, is that the powers of the stellar seeds are communicated along with the stellar light, such that each shares its virtue with the others. They 'conspire', 'inspire', and 'perspire' to create a universal, mutual harmony.

This necessity that the stars exercise their generative powers internally is what led Patrizi to approve of pluralism, the theory advocated by those 'Pythagoreans' who considered every star a world. But in fact, we cannot say that he believed in ET life in the sense used in this discussion. While sympathising with the so-called Pythagorean philosophy, Patrizi hedged his bets and presented the Platonic/Stoic alternative—rather than having their own 'terrestrial' regions, these bodies are living, celestial worlds. Considering his other chapters on the stars and the planets, the Pythagorean theory would not fit with his concept of the planets and stars as bright, flaming condensations of the liquid ether. The two denser compactions of that primeval liquid, the earth and the moon, are separated from the rest by their possession of seas, mountains, and creatures. These two bodies are made a pair, one on each

---

[81] Patrizi 1591, fols 116^{r-v}: 'Stellae, semina quae habent singulae. Vel in se ipsas iaciunt, vel in socias mutuo, vel enim ea. iaciunt, vel non iaciunt. Iacere hoc, foecundare intelligimus. Si non foecundant, ociosa, et semina sunt, ociosae et stellae, et frustra a Conditore conditae sunt. At id dicere est nefarium. Foecundant igitur. Id vero, vel in se ipsis, vel extra se? Et in se et extra se. In se quidem, ut modo quo illis convenit, foecundentur et prolificent. At sensus noster id non cognoscit. Cognoscit tamen ratio. Dictat enim, Deum stellas non frustra fecisse. Frustra autem essent, si operarentur nihil. Semina ergo quaeque sua in se foecundant primo. Atque ita Pythagoreis assentiendum illis, qui singulas stellas singulos esse mundos affirmarunt. Qui suos aeres, suas aquas, suas terras continerent, et animalia in eis habitarent puriora, nostrisque diviniora. Idque longe est probabilius, quam putare, aut asserere, tot, ac tanta, et tam pulchra corpora, vel frustra esse a Conditore fabrecta, aut quasi frustra. Aut si hoc nulla admittit philosophia, necesse est illud admittere, aut aliquid simile, ut scilicet, et ipsa sidera, mundi sint coelestes, spiritu, animo, menteque propriis viventes, et viventes mundi totius vita communi, ex communi tum mente, tum animo, tum spiritu. Extra vero se, agant procul, quam procul lumina sua iacere possunt et cum luminibus etiam seminum vires, quibus sese mutuo foveant, et sibi mutuo faveant, et inter se invicem conspirent, et sibi invicem, et toti, bona inspirent, et bonitate, sibi a Conditore data, omnia perspirent, et in unum ita conspirent, ut harmoniam illam divini artificis, omnia consonant, et personent. Non eam quam fatue in Pythagoreis irrisit Aristoteles, sed eam, quam et ipsi in coelestibus, et nos in universo, dictatu necessariae rationis esse, et asserimus, et contendimus. Nam si stellae in se tantum agere dicerentur, frustra quidem non esse factae dicerentur, sed quasi frustra, si dicerentur, non in se mutuo quicquam agere, et non ope mutua, sese iuvare. Aut si tanta in hylaeum mundum lumina tot numero funderent, quae ad eum vix pertingunt. Aut si solius tanta terrae gratia factae esse crederentur.'

side of the terrestrial/ethereal divide, and that pair is then distinguished from the other celestial bodies. This makes an interesting contrast with Benedetti, who had differentiated between earth and moon, but had then repeated that pattern for the other planets.

Yet regardless of how inconsistent the Pythagorean theory may seem with Patrizi's cosmology, he still described the stars as worlds (*mundi*) in his alternative. They must have, at the very least, some independent purpose and self-contained existence. It is telling that Patrizi, the moon aside, brought up the question of a wider pluralism not in his chapter 'On the planets', or 'Whether the stars are fires', or indeed 'Whether the world is finite', but rather in his chapter on the *effects* of the stars. One could connect Patrizi's tentative pluralism to the size of his universe; one could also connect it to his reduction of the entire universe, and not just the earth, to the same four principle constituents; or, one could connect it to his belief that God's infinite goodness must be manifested in the world. These are undeniably factors, yet Patrizi himself tied it most directly to celestial influence. Influence underpins the symbiotic relationship between the earth and its celestial partner, the moon, and furthermore it determines what is happening in and amongst the stars.

Strangely enough, after stressing so earnestly that the light of the stars barely reaches the earth, and that it is highly unreasonable to think that all these incredible bodies were generated by God 'for the favour of only one, and one such inadequate and deformed body, as the earth', Patrizi presented an alternative theory which rescued the astrological qualities of each and every star and planet.[82] While the earth does not receive much stellar light, and so would not benefit directly from stellar 'seeds', it does, on the other hand, come into closer contact with the sun, which seems to us to be the sole generator of all things. Is it possible, he asked, that the stars project their light, heat and seminal virtues altogether into the sun and the moon, which then administer vicariously to the regions of generation, the air, water and earth (Patrizi 1591, fol. 116ᵛ)? This is reminiscent of Plutarch's description of the moon as a point of confluence, and Cusa's theory that the earth receives the influences of the stars through the medium of the sun and moon.

Patrizi truly seems to represent an intermediary stage. He refashioned the cosmic causal chain to operate both laterally and hierarchically. The stars receive seminal powers from the empyrean heaven, which they utilise within themselves, making each its own living world. At the same time, they act outside of themselves, 'first in the other stars, then in the sun, then in the moon, and then in the material world, the same gifts which come from above being transferred below'.[83] The mutual dimension to Patrizi's theories of celestial influence was crucial to his vision of a living, harmonious and sympathetic world system, where there is no discord 'which does

[82] Patrizi 1591, fol. 116ᵛ: 'Ad quam, lumina ipsarum, aut non, aut vix pertingunt. Ac certe ratio affirmare, aut credere non audet, tam ingentia corpora, terra multo maiora, tot numero, et pulchritudine tanta, fuisse a Conditore Deo, unius tantum, et tam exigui, et tam deformis corporis, qualis terra est, gratia procreata.'

[83] Patrizi 1591, fol. 116ᵛ: 'Agere quoque ea. extra se, tum in astra alia, tum in Solem, tum in Lunam, tum in hyleum, eadem dona, quae sibi a superis venerunt, in infera transfundendo.'

not contribute to the ornament, perfection and conservation of the universe'.[84] While acknowledging his debt to the Platonic, and indeed Hermetic, traditions, we can also, in light of his theories of the rotation of the earth and the matter and motion of the heavens, agree with Patrizi that he was philosophising in a new way.

The salient point to our larger discussion is that the period which saw the dawn of pluralism was a period which saw generation as related to astrological forces. Consider this passage from Patrizi's chapter on the substance of the earth:

> It [the earth, or sediment] is enormously affected by the heavens, that is, by the stars, the planets, the sun, and the moon. It is also affected by the air and the water. From the celestial bodies it receives especially three powers. Two are evident, i.e., light and heat. The third power, however, is less known, and it consists in certain occult influences from celestial and ethereal impressions. Since indeed the earth is inert in itself and in potency, and yet we can see in it certain kinds of works which do not seem to be able to arise from proximate seeds of things, it is necessary that these extraordinary qualities flow in and are poured in from a different source.[85]

Spontaneous generation is an example of the link between the perceived complexity of celestial influence and the relative ignorance of the mechanics of terrestrial generation. For a Platonist with occult leanings like Patrizi, the light and heat which are communicated between celestial bodies are not simple forces, but transmitters of seminal virtues which, given a passive and mutable celestial substance, open the door for pluralist ideas.

2.6 The Astrobiology of Giordano Bruno

No history of pluralism or extraterrestrials can ignore the figure of Giordano Bruno (1548–1600). The wealth of secondary literature on his philosophy and ideas is such that little can be added here. Rather, we can only attempt to highlight certain aspects of his cosmological theories which demonstrate the embedded relationship between celestial influence and pluralism. His name more than any other is attached to the infinity of the universe. He drew inspiration from many of the sources discussed so far, but differed from all. Unlike Lucretius and the atomists, he theorised not a plurality of *kosmoi* but the infinite extent of this one. Unlike Cusa, whose theories aligned quite closely to his own, he didn't advise learned ignorance, or distinguish between an infinite creator and an 'unlimited' creation, but preached his absolute faith in the infinity of the created world. Unlike Patrizi, his infinity was not limited

---

[84] Patrizi 1591, fol. 117^{r}: '...ut nulla in eo sit dissonantia, nulla discordia, quae ad universi, tum ornamentum, tum perfectionem, tum conservationem, non conferat'.

[85] Patrizi 1591, fol. 153^{r}: 'Patitur autem a coelo maxime, ab astris videlicet, a planetis, a sole, a luna; patitur et ab aere, et ab aqua. Ab illis quidem tria maxime, duo indubia, lumen et calorem: tertium autem minus notum, influxus nimirum quosdam occultos, coelestium aetherearumque impressionum. Cum enim ipsa per se iners sit, et inefficax, cernantur autem in ea. opera quaedam, quae a rerum seminibus proximis, oriri non videantur posse, necesse est, aliunde in eam qualitates illas miras influi, et infundi.'

to a heavenly realm or world of pure light, but was filled with an innumerable number of material things:

> For I do not insist on infinite space, nor is Nature endowed with infinite space for the exaltation of size or of corporeal extent, but rather for the exaltation of corporeal natures and species, because infinite perfection is far better presented in innumerable individuals than in those which are numbered and finite.[86]

Thus, Bruno's infinite universe has a certain right to be called original, being unlike those that had come before.[87] His cosmology also argued, significantly, that the stars were other bodies like the sun, orbited by other planets like the earth. This sun/earths system was for him the base unit which he then repeated *ad infinitum*, creating a universe which sat uneasily, according to his critics, between the embodiment of plenitude and the superfluity of infinite repetition.

His own imagined critic, the character of Burchio in his dialogue *De l'infinito universo et mondi* ('On the Infinite Universe and Worlds', 1584), brought up one such objection to this infinity:

> Burchio: 'Where then is that beautiful order, that lovely scale of nature rising from the denser and grosser body which is our earth, to the less dense [sphere] which is water and on to the subtle [sphere] which is vapour, to the yet subtler which is pure air, on to the subtlest which is fire and finally to the divine which is the celestial body? From the obscure to the less obscure, to the brighter and finally to the brightest? From the dark to the most brilliant, from the alterable and corruptible to liberation from all change and corruption? From the heaviest to the heavy, thence to the light, on to the lightest and finally to that which is without weight or lightness? From that which moveth toward the centre to that which moveth from the centre and then to that which moveth around the centre?'
>
> Fracastoro: 'You would like to know where is this order? In the realm of dreams, fantasies, chimeras, delusions.'[88]

Bruno did indeed do away with that ordered structure of the spherical Aristotelian universe which had been maintained to some extent by other early Copernicans. This should and did have profound implications for astrology, a science to which Bruno alternately showed disdain and great interest (Spruit 2002). Yet the Aristotelian system was only one pillar of astrology in the Renaissance and early modern periods, existing alongside Platonic and Stoic concepts of sympathy and

[86] Bruno 1584, 11: '...perché io non richiedo il spacio infinito, e la natura non ha spacio infinito, per la dignità della dimensione o della mole corporea, ma per la dignità delle nature e specie corporee; perché incomparabilmente meglio in innumerabili individui si presenta l'eccellenza infinita, che in quelli che sono numerabili e finiti.' Translated in Bruno 1950, 257.

[87] See Granada 1992; Tessicini 2007.

[88] Bruno 1584, 85–86: '*Burchio*: Ove è dunque quel bell'ordine, quella bella scala della natura, per cui si ascende dal corpo piú denso e crasso, quale è la terra, al men crasso, quale è l'acqua, al suttile, quale è il vapore, al piú suttile, quale è l'aria puro, al suttilissimo, quale è il fuoco, al divino, quale è il corpo celeste? dall'oscuro al men oscuro, al chiaro, al piú chiaro, al chiarissimo? dal tenebroso al lucidissimo, dall'alterabile e corrottibile al libero d'ogni alterazione e corrozione? dal gravissimo al grave, da questo al lieve, dal lieve al levissimo, indi a quel che non è né grave né lieve? dal mobile al mezzo, al mobile dal mezzo, indi al mobile circa il mezzo? *Fracastoro*: Volete saper ove sia questo ordine? Ove son gli sogni, le fantasie, le chimere, le pazzie.' Bruno 1950, 313–14.

determinism. As previously mentioned, the reality of celestial influence was so paradigmatic that it could adapt to varying cosmological assumptions. Bruno's involvement with astrology in practical terms leaned towards the Hermetic and overtly magical, but when it came to the workings of the universe at large he relied on Cusa's theory of interplanetary exchange and a decidedly Stoic plenism (Spruit 2002, 247–49; Westman 2011, 306).

For Bruno's universe was not a collection of separate worlds, but was in itself truly one. As Thomas Aquinas had said, 'the world is called one by the unity of order, according to which certain things are ordered to others'.[89] The basis for this order for Bruno was not a vertical scale or hierarchy, but rather the motion and influence which determined the relationship between individuals. This motion and influence rested in turn upon Bruno's vision of the ether and the World Soul. As the speaker Fracastoro summarises in *De l'infinito*:

> Thus, there is not merely one world, one earth, one sun, but as many worlds as we see bright lights around us, which are neither more nor less in one heaven, one space, one containing sphere than is this our world in one containing universe, one space or one heaven. So that the heaven, the infinitely extending air, though part of the infinite universe, is not therefore a world or part of worlds; but is the womb, the receptacle and field within which they all move and live, grow and render effective the several acts of their vicissitudes; produce, nourish and maintain their inhabitants and animals; and by certain dispositions and orders they minister to higher nature, changing the face of single being through countless subjects.[90]

The heaven is space, is air, is ether, but is not empty void. He argues that space itself is in fact a kind of matter, and that therefore it has both quality and action (Bruno 1950, 272). It is both the infinite space where motion takes place, and also the source of all motive power, as it not only surrounds bodies but penetrates them and is incorporated into their being, giving each the ability to generate motion (Bruno 1950, 273). More than this, the ether and the universe as an infinite space is in itself an animal, for 'it is imbued with all soul and embraceth all life and it is the whole of life' (Bruno 1950, 300).

Yet apart from this living nature, motive power, and generic quality, the ether can have no specific quality of its own, as one of its main tasks is as a neutral carrier and medium for the influences of the bodies within it (Bruno 1950, 372). It is here that the Stoic plenum met Platonic sympathy and antipathy, boiled down to an opposi-

---

[89] Thomas Aquinas, *Summa theologiae*, I, q. 47, a. 3: 'Mundus enim iste unus dicitur unitate ordinis, secundum quod quaedam ad alia ordinantur.'

[90] Bruno 1584, 98–99: 'Di maniera che non è un sol mondo, una sola terra, un solo sole; ma tanti son mondi, quante veggiamo circa di noi lampade luminose, le quali non sono piú né meno in un cielo ed. un loco ed. un comprendente, che questo mondo, in cui siamo noi, è in un comprendente, luogo e cielo. Sí che il cielo, l'aria infinito, immenso, benché sia parte de l'universo infinito, non è però mondo, né parte di mondi; ma seno, ricetto e campo in cui quelli sono, si muoveno, viveno, vegetano e poneno in effetto gli atti de le loro vicissitudini, producono, pascono, ripascono e mantieneno gli loro abitatori ed. animali, e con certe disposizioni ed. ordini amministrano alla natura superiore, cangiando il volto di uno ente in innumerabili suggetti.' Bruno 1950, 322–23.

tion of hot and cold reminiscent of the philosophy of Archelaus (fifth century BCE).[91] This combination was the key to Bruno's new order:

> We shall understand that the orbs and spheres of the universe are not disposed one beyond another, each smaller unit being contained within a greater—as, for example, the infoldings of an onion. But throughout the ethereal field, heat and cold, diffused from the bodies wherein they predominate, gradually mingle and temper one another to varied degrees, so as to become the proximate origin of the innumerable forms and species of being.[92]

The two types of primary bodies in the universe, hot suns and cold, watery earths, exist in a symbiotic relationship, each being unable to endure without the other. 'This shows', he said in the philosophical poem *De innumerabilibus, immenso et infigurabili* ('On Innumerable Things, Immensity and Unrepresentable Reality', 1591), 'that nature provides for the motion, generation and continued existence of things by the concourse of diverse and opposite entities'.[93] The ether is the 'womb' of the universe, in which all the spheres move and effect the changes within themselves, producing and maintaining populations of inhabitants and animals. Each globe itself, and indeed the entire universe, should be considered an animal (Bruno 1950, 300). This then took on an explanatory role in terms of celestial ordering and motion. They can both be explained through a biological imperative, such that each globe with its inhabitants is preserved by the mediation of celestial influence which ensures optimal living conditions.

Just as the earth is heated and lit by the sun, allowing things to live here, so is the sun chilled and watered by its surrounding planets, allowing things to live there (Bruno 1950, 309). This influence of the earth was explained not only in terms of a moistening effect, as we had seen in Plutarch and Cusa, but also in line with Bruno's atomistic philosophy:

> Nevertheless, the universe being infinite, and the bodies thereof transmutable, all are therefore constantly dispersed and constantly reassembled; they send forth their substance, and receive within themselves wandering substance. Nor doth it appear to me absurd or inconvenient, but on the contrary most fitting and natural that finite transmutations may occur to a subject; wherefore particles of [elemental] earth may wander through the ethereal region and may traverse vast space now to this body, now to that, just as we see the same particles change their position, their disposition and their form, even when they are yet close to us.[94]

---

[91] On Archelaus, see Laertius 1959, I, 144–49.

[92] Bruno 1584, 106: 'Comprenderemo, che non son disposti gli orbi e sfere nell'universo, come vegnano a comprendersi l'un l'altro, sempre oltre ed. oltre essendo contenuto il minore dal maggiore, per esempio, gli squogli in ciascuna cipolla; ma che per l'etereo campo il caldo ed. il freddo, diffuso da' corpi principalmente tali, vegnano talmente a contemperarsi secondo diversi gradi insieme, che si fanno prossimo principio di tante forme e specie di ente.' Bruno 1950, 328.

[93] Bruno 1879, I, 213: 'Unum primorum corporum genus absque alio consistere minime posse, illud indicat, quod diversorum oppositorumque concursu ad motum, ad generationem et rerum consistentiam natura provideat.'

[94] Bruno 1584, 47–48: 'Tutta volta, essendo l'universo infinito e gli corpi suoi tutti trasmutabili, tutti per conseguenza diffondeno sempre da sé e sempre in sé accoglieno, mandano del proprio fuora e accogliono dentro del peregrino. Non stimo che sia cosa assorda ed. inconveniente, anzi convenientissima e naturale, che sieno transmutazion finite possibili ad accadere ad un soggetto; e però de particole de la terra vagar l'eterea regione e occorrere per l'inmenso spacio ora ad un corpo

The continual concourse of atoms which is the basis of all change, generation, and corruption, takes place not only on a local but also on an interplanetary and interstellar scale. The physical grounding of astrology could adapt, it seems, both to Bruno's reimagining of space and cosmic order as well as to his atomistic theory of matter. It formed an essential part of his vision of a universe operating according to a quasi-mechanical vitality.

When Bruno argued specifically for the inhabitation of all the celestial bodies—for which he is honoured as the founding father of modern pluralism—his words echoed those of Cusa. All these celestial bodies 'are either themselves suns, or the sun doth diffuse to them no less than to us those most divine and fertilizing rays' (Bruno 1950, 323). This influence alone is enough to guarantee the existence of inhabitants, as elsewhere he argued that generation by reproduction was not a universal truth. The 'mere force and innate vigour of nature' is enough to bring forth many and varied animals without any 'original act of generation' (Bruno 1950, 376). In the dedicatory letter to *De immenso*, again echoing Cusa, Bruno described the perfection of the universe in a series of three-part principles, one of which involved the efficacy of active power, the disposition of passive power, and the dignity of the effects (Bruno 1879, I, 198). His physical universe encapsulated this with the generation of living things by the action of celestial influence on the passive material of the celestial bodies.

Bruno's theories on the motion of these bodies were of course animistic. The living planets, endowed with sensitive and intellective souls, carry out vagaries of motion which are to be understood not according to geometrical laws but rather in terms of purpose and desire (McMullin 1987, 61). These earths move so as to perpetuate and renew not only their life, but that of their inhabitants, and this motion is oriented not only to the sun but also to the other planets which share vital principles among themselves.[95] This purposive motion is not without a certain amount of providential design, however. There could be a danger that contrary bodies placed too close together would be harmful rather than beneficial, and as well similar bodies could block one another from receiving necessary dissimilar qualities. The truth of this is demonstrated by eclipses, 'when our frailty suffereth considerable damage through the interposition between ourselves and the sun of that other earth that we name the moon' (Bruno 1950, 335). It seems that Providence determined certain cosmic distances, and then provided the celestial bodies with the intellective ability to decide the most productive course. This was an animistic physics built upon principles of analogy: on the one hand, from motion as witnessed in living creatures to motion in general; on the other, from the earth as an individual to the earth as a generic type of celestial body.

This analogy, however, ran into obvious difficulties because he had no real knowledge about any of these bodies except for the one which was the 'prototype of the class' (Breiner 1978, 28). In the *Ash Wednesday Supper*, and also in *De immenso*,

ora ad un altro, non meno che veggiamo le medesime particole cangiarsi di luogo, di disposizione e di forma, essendono ancora appresso di noi.' Bruno 1950, 284–85.

[95] See Michel 1973; Bruno 1975, 91, 156, 185, 213; McMullin 1987, 216.

we see how this led him into problems. Just like his contemporary Benedetti, Bruno misunderstood Copernican astronomy and distorted it according to his own understanding of the nature of celestial bodies and their relationships. He suggested that the moon does not orbit around the earth as its central point, but rather that they are positioned on opposite sides of a single epicycle with a common centre, and in this arrangement they orbit around the sun together (Bruno 1975, 192). Not only this, but he stated that this situation is replicated in the case of Mercury and Venus, meaning they orbit around each other and share a common orbit around the sun. All of this contravened the existing body of observational evidence (Bruno 1879, I, 397; McMullin 1987, 56–59). So we see in Bruno another case of the moon/earth pairing being extrapolated to cover the other celestial bodies, but in his cosmology the idea of the moon as simply an auxiliary support mechanism did not make sense. It must be a living and inhabited body in its own right, on equal footing to the earth, otherwise the principle of universal homogeneity would be undermined.

This brief survey of Bruno's cosmology is intended to demonstrate his participation in a broader movement away from a top-down theory of celestial causation to a belief in the mutual influence of celestial bodies, and the stimulus this move gave to pluralist philosophies. How then does this change our perception of Bruno in the history of the ET debate, and the history of science more generally? He has, it has been said, assumed a 'rather awkward niche' in the history of science (Breiner 1978, 35). His scepticism of the ability to predict planetary motion, and his reliance on philosophical theory rather than observation meant that he was quite quickly superseded by Tycho Brahe, Kepler and Galileo in terms of astronomy and physics. The influence he had on English thinkers during his time there in England has been downplayed in a recent article by Feingold, who suggested that Bruno gained more from the exchange than vice versa (Feingold 2004, 329–46). Many authors downplay his contributions to either Copernicanism or science in general (McMullin 1987, 74). Koyré regretfully concluded that he was 'not a very good philosopher' (Koyré 1957, 54). He was a Hermetic *magus*, a 'benign high priest', with an outlook more ancient than modern (Yates 1964; Candela 1998, 351).

Yet there is no reason to necessarily remove him from his exalted place in the history of pluralism, extraterrestrials, or astrobiology. He was its loudest voice at the time, evangelically spreading the word of an infinitely populated universe, proving himself an original and influential thinker. Though for a long time the notion that he was executed for specifically this belief was rejected, a new article by Martinez, based on a study of the available inquisitorial records, suggests that perhaps it was just that (Martinez 2016). And perhaps he wasn't the only one, as a report from February 12, 1600, a few days before Bruno's execution, makes reference to another 'curate' burned at the stake for believing in 'so many suns' (Martinez 2016, 365).[96] Bruno's place in the history of pluralism need not be diminished, then, but rather

[96] Martinez suggests that the curate may have been Friar Celestino of Verona, born Giovanni Antonio Arrigoni. The document in question is: Report [12 February 1600, Rome], Archivo Orsini. I. Corrispondenza, b. 380, n. 385, Archivo Storico Capitolino, Roma. On the relationship between Bruno and Fra Celestino, see now Maifreda 2016.

developed. Lovejoy's thesis that pluralism was stimulated less by Copernicus and more by a resurgent Platonism has some merit, but needs to be expanded. It was not the principle of plenitude alone which opened the door, but also the growing emphasis on the Platonic and Stoic dimensions of astrological theory, involving living and interacting celestial bodies. Bruno should be remembered not only for enlarging and populating the universe, but also for his theories on celestial influence and motion, and the ties that bound all these together.

The history of astrobiology may be ideally suited for this re-evaluation. In his 1962 work *La Cosmologie de Giordano Bruno*, Paul-Henri Michel twice used the term *astrobiologie*:

> The description of the planetary system no longer depends on cosmic geometry but on astrobiology which makes the motion of celestial bodies a function of their nature; and if the irregularity of this motion reveals the imperfection of the tangible being, it is equally a sign of its freedom (Michel 1973, 216).

And again:

> In Bruno's mind, the life of celestial bodies is not a legend, a symbol or a metaphor, but one of the theorems of his physics and the fundamental axiom of his astrobiology (Michel 1973, 300).

Michel was referring not so much to Bruno's thoughts about life *on* the other globes, but rather to the life of the globes themselves, and to the biological astronomy that Bruno expounds. It could be argued that this is simply a now-defunct usage of the term. On the other hand, it could be an opportunity to enrich our understanding of the history of astrobiology. In the Renaissance, astrology and celestial animism were the equivalent interdisciplinary fields melding astronomy and biology, and the ways that people understood life in its cosmic context.

2.7 Conclusion

This chapter has touched on a broad range of philosophical, scientific and theological topics over an even broader time frame. The treatment of these subjects has been necessarily cursory, but hopefully certain themes and trends have been sufficiently illustrated. Plutarch's *De facie* preserved Platonic and Stoic approaches to the question of the earth's relationship to the moon and the rest of the celestial realm. He discussed the relativity of cosmological perspective, the influence of the earth on other bodies, and the comparative teleological implications of influence and inhabitation in the case of the moon. It was influence which dominated the teleological discussion of the heavens in the Middle Ages, with an astrologised version of Aristotle's biology being transmitted via Arabic philosophy, and the ordered celestial hierarchy standing as both sign and causal agent of God's providence.

Cusa, Palingenius, Digges, Benedetti, Patrizi and Bruno are examples of how Platonic philosophy and cosmology, absorbed in different forms through classical and medieval philosophy, fused with the medieval and Renaissance understanding of that universal astrological law of nature in such a way as to facilitate the rehabilitation of the belief in a plurality of worlds. The introduction of Copernicanism and the growing criticism of the Aristotelian/Ptolemaic view of the cosmos did not put an end to that law, as the next two chapters of this paper will demonstrate. The belief in ET life, which would eventually replace astrological thinking as a teleological explanation for the creation of the planets and stars, matured in the seventeenth century, in an environment in which astrology was still considered to play an important role in generation and corruption, and therefore in the process of life.

References

Achtner, Wolfgang. 2005. Infinity in science and religion: The creative role of thinking about infinity. *Neue Zeitschrift für Systematische Theologie und Religionsphilosophie* 47: 392–411.

Al-Kindi. 1974. *De radiis*. Edited by Marie-Thérèse d'Alverny and Françoise Hudry. Archives d'histoire doctrinale et littéraire du Moyen Âge 41: 139–260.

Aristotle. 1943. *Generation of Animals*. Trans. A.L. Peck. Cambridge, MA: Harvard University Press.

Bacchelli, Franco. 1999. Palingenio e la crisi dell'aristotelismo. In *Sciences et religions de Copernic à Galilée (1540–1610)*, ed. Catherine Brice and Antonella Romano, 357–374. Rome: École française de Rome.

———. 2001. Palingenio Stellato e la sua fortuna europea. In *Napoli Viceregno spagnolo: una capitale della cultura alle origini dell'Europa moderna (sec. XVI–XVII)*, ed. Monika Bosse and André Stoll, 153–166. Naples: Vivarium.

Barnes, Robin B. 2015. *Astrology and reformation*. New York: Oxford University Press.

Benedetti, Giovanni Battista. 1585. *Diversarum speculationum mathematicarum et physicarum liber*. Turin: Heirs of Nicola Bevilaqua.

Bertolacci, Amos. 2013. Averroes against Avicenna on human spontaneous generation: The starting-point of a lasting debate. In *Renaissance Averroism and its aftermath: Arabic philosophy in early modern Europe*, ed. Anna Akasoy and Guido Giglioni, 37–54. Dordrecht: Springer.

Blair, Ann. 1990. Tycho Brahe's critique of Copernicus and the Copernican system. *Journal of the History of Ideas* 51: 355–377.

Breiner, Laurence A. 1978. Analogical argument in Bruno's *De l'infinito*. *MLN* 93: 22–35.

Brickman, Benjamin. 1941. *An introduction to Francesco Patrizi's Nova de universis philosophia*. New York: Columbia University.

Brosseder, Claudia. 2005. The writing in the Wittenberg sky: Astrology in sixteenth-century Germany. *Journal of the History of Ideas* 66: 557–576.

Bruno, Giordano. 1584. *De l'infinito universo et mondi*. Stampato in Venetia [i.e. London]: J. Charlewood.

———. 1879. *Opera Latine conscripta*. Edited by Francesco Fiorentino. Vol. 1, 8 vols. Naples/Florence: Morano and Le Monnier.

———. 1950. *On the Infinite Universe and Worlds*. Trans. Dorothea Waley Singer in ead. *Giordano Bruno: His life and thought*. New York: Schuman.

———. 1975. *The Ash Wednesday supper = La cena de le Ceneri*. Trans. Stanley L. Jaki. The Hague: Mouton.

Burnett, Charles. 2009. *Arabic into Latin in the Middle Ages: The translators and their intellectual and social context*. Farnham: Ashgate.

Campion, Nicholas. 2008. *The dawn of astrology: A cultural history of Western astrology*. London: Continuum.

———. 2009. *A history of Western astrology volume II: The medieval and modern worlds.* London: Bloomsbury.
Candela, Giuseppe. 1998. An overview of the cosmology, religion and philosophical universe of Giordano Bruno. *Italica* 75: 348–364.
Cardano, Girolamo. 1554. *In Cl. Ptolemaei IIII de astrorum iudiciis aut quadripartitae constructionis libros commentaria.* Basel: Henricus Petrus.
Chlup, Radek. 2000. Plutarch's dualism and the Delphic cult. *Phronesis* 45: 138–158.
Cicero. 1896. *On the Nature of the Gods.* Trans. Francis Brooks. London: Methuen.
Coones, Paul. 1983. The geographical significance of Plutarch's dialogue, *Concerning the face which appears in the orb of the moon. Transactions of the Institute of British Geographers* 8: 361–372.
Copenhaver, Brian P. 1984. Scholastic philosophy and renaissance magic in the *De vita* of Marsilio Ficino. *Renaissance Quarterly* 37: 523–554.
Cusa, Nicholas of. 1932. *Opera omnia.* Vol. 1. Leipzig: Felix Meiner.
———. 1985. *On Learned Ignorance: A Translation and an Appraisal of* De docta ignorantia. Trans. Jasper Hopkins. 2nd. Minneapolis: A.J. Benning Press.
———. 1997. *Selected spiritual writings*, ed. H. Lawrence Bond. Mahwah: Paulist Press.
Dales, Richard C. 1980. The de-animation of the heavens in the Middle Ages. *Journal of the History of Ideas* 41: 531–550.
Daneau, Lambert. 1578. *The Wonderfull Woorkmanship of the World.* Trans. Thomas Twyne. London: Andrew Maunsell.
Davidson, Herbert A. 1992. *Alfarabi, Avicenna, and Averroes, on intellect: Their cosmologies, theories of the active intellect, and theories of human intellect.* New York: Oxford University Press.
Dick, Steven J. 1982. *Plurality of worlds: The origins of the extraterrestrial life debate from Democritus to Kant.* Cambridge/New York: Cambridge University Press.
Dick, Steven J., and James Edgar Strick. 2005. *The living universe: NASA and the development of astrobiology.* New Brunswick: Rutgers University Press.
Digges, Leonard. 1576. *A prognostication euerlastinge of right good effecte.* London: Thomas Marsh.
Drake, Stillman. 1970. Giovanni Battista Benedetti. In *Dictionary of scientific biography*, ed. C.C. Gillispie, vol. 1, 604–609. New York: Charles Scribner's Sons.
Duhem, Pierre Maurice Marie. 1985. *Medieval Cosmology: Theories of Infinity, Place, Time, Void, and the Plurality of Worlds.* Trans. Roger Ariew. Chicago: University of Chicago Press.
Ernst, Germana. 1991. Astrology, religion and politics in Counter-Reformation Rome. In *Science, culture and popular belief in Renaissance Europe*, ed. Stephen Pumfrey, Maurice Slawinski, and Paolo L. Rossi, 249–273. Manchester: Manchester University Press.
———. 2001. 'Veritatis amor dulcissimus': aspects of Cardano's astrology. In *Secrets of nature: Astrology and alchemy in early modern Europe*, ed. William Royall Newman and Anthony Grafton, 39–68. Cambridge, MA/London: MIT Press.
Fabbri, Natacha. 2012. The Moon as another Earth: What Galileo owes to Plutarch. *Galilaeana* 9: 103–105.
———. 2016. Looking at an earth-like moon and living on a moon-like earth in Renaissance and early modern thought. In *Early modern philosophers and the Renaissance legacy*, ed. Cecilia Muratori and Gianni Paganini, 135–151. Dordrecht: Springer.
Feingold, Mordechai. 2004. Giordano Bruno in England, revisited. *Huntington Library Quarterly* 67: 329–346.
Ficino, Marsilio. 2001. *Platonic theology*, ed. Michael J.B. Allen and James Hankins. Vol. 1. 6 vols. Cambridge, MA: Harvard University Press.
Freudenthal, Gad. 2002. The medieval astrologization of Aristotle's biology: Averroes on the role of the celestial bodies in the generation of animate beings. *Arabic Sciences and Philosophy* 12: 111–137.

———. 2009. The astrologization of the Aristotelian cosmos: Celestial influences on the sublunary world in Aristotle, Alexander of Aphrodisias, and Averroes. In *New perspectives on Aristotle's* De caelo, ed. Alan C. Bowen and Christian Wildberg, 239–281. Leiden: Brill.

Funkenstein, Amos. 1986. *Theology and the scientific imagination from the Middle Ages to the seventeenth century*. Princeton: Princeton University Press.

Gotthelf, Allan. 2012. *Teleology, first principles, and scientific method in Aristotle's biology*. Oxford: Oxford University Press.

Granada, Miguel A. 1992. Bruno, Digges, Palingenio: omogeneita ed eterogeneita nella concezione dell'universo infinito. *Rivista di storia della filosofia* 47: 47–74.

Grant, Edward. 1981. *Much ado about nothing: Theories of space and vacuum from the Middle Ages to the scientific revolution*. Cambridge/New York: Cambridge University Press.

———. 1994. *Planets, stars, and orbs: The medieval cosmos*, 1200–1687. Cambridge: Cambridge University Press.

Hardy, Nicholas. 2015. Is the *De rerum natura* a work of natural theology? Some ancient, modern, and early modern perspectives. In *Lucretius and the early modern*, ed. David Norbrook, Stephen Harrison, and Philip Hardie, 200–221. Oxford: Oxford University Press.

Haskell, Yasmin. 2015. Poetic flights or retreats? Latin Lucretian poems in sixteenth-century Italy. In *Lucretius and the early modern*, ed. David Norbrook, Stephen Harrison, and Philip Hardie, 91–122. Oxford: Oxford University Press.

Henry, John. 1979. Francesco Patrizi da Cherso's concept of space and its later influence. *Annals of Science* 36: 549–573.

———. 2001. Void space, mathematical realism, and Francesco Patrizi da Cherso's use of atomistic arguments. In *Late medieval and early modern corpuscular matter theories*, ed. Christoph Herbert Lüthy, John Emery Murdoch, and William Royall Newman, 133–161. Leiden: Brill.

Hershbell, Jackson P. 1982. Plutarch and Democritus. *Quaderni urbinati di cultura classica* 10: 81–111.

Hirai, Hiro. 2008. Earth's soul and spontaneous generation: Fortunio Liceti's criticism of Ficino's ideas on the origin of life. In *Platonism at the origins of modernity: Studies on Platonism and early modern philosophy*, ed. Douglas Hedley and Sarah Hutton, 273–299. Dordrecht: Springer.

———. 2011. *Medical humanism and natural philosophy: Renaissance debates on matter, life and the soul*. Leiden: Brill.

Jaki, Stanley L. 1978. *Planets and planetarians: A history of theories of the origin of planetary systems*. Edinburgh: Scottish Academic Press.

Johnson, Francis R., and Sanford V. Larkey. 1934. Thomas Digges, the Copernican system, and the idea of the infinity of the universe in 1576. *The Huntington Library Bulletin* (5): 69–117.

Jouanna, Jacques. 2001. *Hippocrates*. Trans. M.B. DeBevoise. Baltimore: Johns Hopkins University Press.

Kant, Immanuel. 1981. *Universal Natural History and Theory of the Heavens*. Trans. Stanley L. Jaki. Edinburgh: Scottish Academic Press.

Kassell, Lauren. 2005. *Medicine and magic in Elizabethan London: Simon Forman, astrologer, alchemist, and physician*. Oxford: Clarendon Press.

Knox, Dilwyn. 2005. Copernicus's doctrine of gravity and the natural circular motion of the elements. *Journal of the Warburg and Courtauld Institutes* 68: 157–211.

Koyré, Alexandre. 1957. *From the closed world to the infinite universe*. Baltimore: Johns Hopkins Press.

Kristeller, Paul Oskar. 1964. *Eight philosophers of the Italian renaissance*. Stanford: Stanford University Press.

Kruk, Remke. 1990. A frothy bubble: Spontaneous generation in the medieval Islamic tradition. *Journal of Semitic Studies* 35: 265–282.

Kusukawa, Sachiko. 1995. *The transformation of natural philosophy: The case of Philip Melanchthon*. Cambridge/New York: Cambridge University Press.

Laertius, Diogenes. 1959. *Lives of eminent philosophers*, ed. R. D. Hicks, 2 vols. Cambridge, MA/London: Cambridge University Press and Heinemann.

Lennox, James. 1982. Teleology, chance, and Aristotle's theory of spontaneous generation. *Journal of the History of Philosophy* 20: 219–239.
Lindberg, David C. 1986. The genesis of Kepler's theory of light: Light metaphysics from Plotinus to Kepler. *Osiris* 2: 4–42.
Lovejoy, Arthur O. 1948 [1936]. The great chain of being: A study of the history of an idea. Cambridge, MA: Harvard University Press.
Lucretius. 2004. *On the Nature of Things*. Trans. William Ellery Leonard. Mineola: Dover.
Maifreda, Germano. 2016. *Giordano Bruno e Celestino da Verona. Un incontro fatale*. Pisa: Edizioni della Normale.
Martinez, Alberto A. 2016. Giordano Bruno and the heresy of many worlds. *Annals of Science* 73: 345–374.
McColley, Grant. 1936. The seventeenth-century doctrine of a plurality of worlds. *Annals of Science* 1: 385–430.
McMullin, Ernan. 1987. Bruno and Copernicus. *Isis* 78: 55–74.
Melanchthon, Philip. 1550. *Initia doctrina physicae*. Wittenberg: Johannes Lufft.
Methuen, Charlotte. 1996. The role of the heavens in the thought of Philip Melanchthon. *Journal of the History of Ideas* 57: 385–403.
Michel, Paul-Henri. 1973. *The Cosmology of Giordano Bruno*. Trans. R.E.W. Maddison. London: Methuen.
Mosley, Adam. 2007. *Bearing the heavens: Tycho Brahe and the astronomical community of the late sixteenth century*. Cambridge: Cambridge University Press.
Muccillo, Maria. 1986. Marsilio Ficino e Francesco Patrizi da Cherso. In Marsilio Ficino e il ritorno di Platone: studi e documenti, ed. Gian Carlo Garfagnini, II, 615–679. Florence: Olschki.
Oestmann, Günther, H. Darrel Rutkin, and Kocku von Stuckrad, ed. 2006. *Horoscopes and public spheres: Essays on the history of astrology*. Berlin/New York: Walter de Gruyter.
Omodeo, Pietro Daniel. 2014. *Copernicus in the cultural debates of the renaissance: Reception, legacy, transformation*. Leiden: Brill.
Palingenius, Marcellus. 1947 [1560]. *The Zodiake of Life*. Trans. Barnabe Googe. Delmar: Scholars' Facsimiles & Reprints.
———. 1996. *Le zodiaque de la vie (Zodiacus vitae): XII livres*. Trans. Jacques Chomarat. Geneva: Librairie Droz.
Patrizi, Francesco. 1591. *Nova de universis philosophia*. Ferrara: Benedetto Mammarello.
Pennuto, Concetta. 2008. The debate on critical days in Renaissance Italy. In *Astro-medicine: Astrology and medicine, East and West*, ed. Anna Akasoy, Charles S. F. Burnett, and Ronit Yoeli-Tlalim, 75–98. Florence.: Sismel – Edizioni del Galluzzo.
Plotinus. 1580. *Opera philosophica omnia*. Trans. Marsilio Ficino. Basel: Pietro Perna.
Plutarch. 1936. Isis and Osiris. In *Moralia*. Trans. F.C Babbit, V, 6–191. Cambridge, MA/London: Harvard University Press and Heinemann.
———. 1957. Concerning the face which appears in the orb of the Moon. In *Moralia*. Trans. W.C. Helmbold and H.F. Cherniss, XII, 34–226. Cambridge, MA/London: Harvard University Press and Heinemann.
Porter, James I. 2016. *The sublime in antiquity*. Cambridge: Cambridge University Press.
Roos, Anna Marie Eleanor. 2001. *Luminaries in the natural world: The sun and the moon in England, 1400–1720*. New York/Oxford: Peter Lang.
Rosen, Edward. 1984. Francesco Patrizi and the celestial spheres. *Physis* 26: 305–324.
Roth, Ulli. 2001a. Die astronomisch-astrologische 'Weltgeschichte' des Nikolaus von Kues im Codex Cusanus 212. Einleitung. *Mitteilungen und Forschungsbeiträge der Cusanus-Gesellschaft* 27: 1–29.
———. 2001b. Cusanus' Weltgeschichte im Codex 212. *Astronomisch-astrologische Überlegungen. Litterae Cusanae. Informationen der Cusanus-Gesellschaft* 1: 2–14.
Rutkin, H. Darrel. 2002. Astrology, natural philosophy, and the history of science, c. 1250–1700: Studies toward an interpretation of Giovanni Pico della Mirandola's *Disputationes adversus astrologiam divinatricem*. PhD Dissertation, University of Indiana.

———. 2013. Astrology and magic. In *A companion to Albert the Great: Theology, philosophy, and the sciences*, ed. Irven Resnick, 451–505. Leiden: Brill.
———. 2018. How to accurately account for astrology's marginalization in the history of science and culture: The central importance of an interpretive framework. *Early Science and Medicine* 23: 217–243.
Saif, Liana. 2015. *The Arabic influences on early modern occult philosophy*. Houndmills/Basingstoke/New York: Palgrave Macmillan.
Siraisi, Nancy G. 1973. *Arts and sciences at Padua: The studium of Padua before 1350*. Toronto: Pontifical Institute of Mediaeval Studies.
Spruit, Leen. 2002. Giordano Bruno and astrology. In *Giordano Bruno: Philosopher of the renaissance*, ed. Hilary Gatti, 229–249. London: Ashgate.
Stuckrad, Kocku von. 2016. Astrology. In A companion to science, technology, and medicine in ancient Greece and Rome, ed. Georgia L. Irby-Massie, I, 114–129. Chichester: Wiley Blackwell.
Świeżyński, Adam. 2016. Where/when/how did life begin? A philosophical key for systematizing theories on the origin of life. *International Journal of Astrobiology* 15: 291–299.
Tessicini, Dario. 2007. *I dintorni dell'infinito: Giordano Bruno e l'astronomia del Cinquecento*. Pisa and Rome: Serra.
Vasoli, Cesare. 1989. *Francesco Patrizi da Cherso*. Rome: Bulzoni.
———. 2002. Sophismata putida: la critica patriziana all dottrina peripatetica dell'eternità e immutabilità del cielo. In *Francesco Patrizi filosofo platonico nel crepuscolo del Rinascimento*, ed. Patrizia Castelli, 167–180. Florence: Olschki.
Vescovini, Graziella Federici. 2014. The theological debate. In *A companion to astrology in the renaissance*, ed. Brendan Maurice Dooley, 99–140. Leiden: Brill.
Volk, Katharina. 2009. *Manilius and his intellectual background*. Oxford: Oxford University Press.
Warren, James. 2004. Ancient atomists on the plurality of worlds. *The Classical Quarterly* 54: 354–365.
Watanabe, Morimichi, Gerald Christianson, and Thomas M. Izbicki, eds. 2013. *Nicholas of Cusa: A companion to his life and his times*. Farnham: Ashgate Publishing Ltd.
Watson, Foster. 1908. *The Zodiacus vitae of Marcellus Palingenius Stellatus: An old school-book*. London: P. Wellby.
Westman, Robert S. 1975. The Melanchthon circle, Rheticus, and the Wittenberg interpretation of the Copernican theory. *Isis* 66: 165–193.
———. 2011. *The Copernican question: Prognostication, skepticism, and celestial order. Berkeley*. London: University of California Press.
Yates, Frances Amelia. 1964. Giordano Bruno and the hermetic tradition. London: Routledge and Kegan Paul.

Chapter 3
William Gilbert: Magnetism as Astrological Influence, and the Unification of the Terrestrial and Celestial Realms

The body of the earth is one and uniform. However, it is not simple in its qualities (as the opinion of many conceives them). A great variety of things blossom on its surface, stirred up both by its internal forces and by the lights of the heavenly bodies ...

But it is only on the surfaces of the globes that the multitude of living and animated beings is clearly perceived, by which great and delightful variety the great maker is well pleased.

Abstract This chapter presents a case study of the natural philosophy of William Gilbert. While saying little about the practice of astrology or the question of extra-terrestrial life *per se*, Gilbert's cosmology was instrumental in the dissolution of the Aristotelian sublunary/celestial divide, and in the application of terrestrial analogies to questions of celestial physics. As well as Gilbert's one published work, *De magnete* (1600), particular attention is paid to his posthumously published *De mundo nostro sublunari philosophia nova* (1651). The chapter describes and analyses Gilbert's philosophy in both its terrestrial and celestial dimensions. Celestial influence is shown to be a major component of Gilbert's system. Gilbert's theories of magnetism and electricity are also examined, especially in relation to cosmological matters. While he only makes passing reference to a variety of living and animated beings on the surfaces of the celestial globes (plural), the assumption of life on the other celestial bodies in fact plays an important role in his rationale for celestial motion.

Gilbert 1651, 108: '…unum enim telluris est corpus et uniforme, non non tamen qualitatibus, ut multorum eas opinio concipit, simplex… et insitis viribus, et luminibus astrorum concussum in eminentiis, multa rerum varietate efflorescat…'

Gilbert 1600, 210: 'Sed in globorum extremitatibus tantum, animarum et animatorum frequentia manifestius cernitur, in quibus summus opifex, maiore et iucunda varietate sibi perplacet.'

© Springer Nature Switzerland AG 2019

J. E. Christie, *From Influence to Inhabitation*, International Archives of the History of Ideas Archives internationales d'histoire des idées 228,
https://doi.org/10.1007/978-3-030-22169-0_3

Keywords William Gilbert · Astrology · Extraterrestrial life · *De mundo* · Magnetism · Animism

3.1 Introduction

As the preceding chapter has hopefully demonstrated, the shift of focus within Western astronomical cosmology from influence to inhabitation was not a clean break or sudden change, nor was it a simple replacement of an occult mentality with scientific speculation. Rather it was a process; one involving a combination of subtle yet profound adjustments in natural philosophy and theology. Having thus far attempted a brief sketch of this process up to the death of Bruno in 1600, this next chapter will present a case study of a philosopher whose major work was published in the same year. William Gilbert (1544–1603), although well-known to the history of science, has little place in the histories of astrology or the ET life debate, yet he introduced and developed theories that are essential to both. He was familiar with the ideas of Cusa, Benedetti, Patrizi and Bruno, and his ideas were in turn highly influential in seventeenth-century physics and cosmology. The analysis of his work presented here is an attempt to both highlight and clarify certain aspects of his somewhat problematic yet highly interesting philosophy, as well as to take a more in depth look at how astrological and pluralist ideas could interact within a single worldview.

In Gilbert's new natural philosophy (*physiologia nova*), we find the possibility of ET life appearing within a universe pervaded by a distinctly novel form of astrological influence. One of the main ambitions of his attack on Aristotle and the Peripatetics was to unify the terrestrial and celestial realms. This involved separating the earth from the planets physically, by the interposition of a vacuum, while bonding them qualitatively. These moves towards a more modern understanding of the heavens were not in themselves inhibitive towards astrology, as Gilbert will demonstrate. Instead, the opening of new lines of analogical reasoning led to a re-appraisal of astrological physics, in which planetary motion was related to celestial influence, and the earth along with the rest of the celestial bodies partook in a mutual exchange of forces. At the same time, the planets became more earth-like, acquiring the crucial attribute of habitability.

The main obstacle facing any historian wishing to study the Elizabethan physician William Gilbert is that, apart from his two published works, *De magnete* (1600) and *De mundo* (1651), there is an almost complete dearth of auxiliary source material. Apart from one letter to William Barlow, which was printed as an appendix to Barlow's *Magneticall Advertisments* (Barlow 1616, 87–88), no other written material by Gilbert survives. Historians, however, have managed to piece together a rough biography from institutional records and other sources. He was born in the town of Colchester, Essex, in or around 1544, and later studied at Cambridge. After a brief spell as senior bursar there in 1570, he left Cambridge to pursue a medical career in London. What followed was an apparently successful practice as a physician spanning just over 30 years, leading to his election as president of the College of Physicians in 1600. He acquired the residence of Wingfield House near St Paul's Cathedral sometime in the mid-1590s, where he lived until moving to court upon his selection as a Royal Physician in 1601, obtaining an appointment for life as physician to Elizabeth I. He was re-appointed in the same post to James I, but died shortly afterwards, the year being 1603 (Pumfrey 2004).

While Gilbert's professional career was in medicine, his intellectual interests encapsulated a much broader swathe of natural philosophy. His main contribution to the field was his *De magnete, magnetisque corporibus, et de magno magnete tellure; physiologia nova, plurimis et argumentis, et experimentis demonstrata* ('On the Magnet, Magnetic Bodies and the Earth, the Great Magnet: A New Philosophy of Nature Demonstrated through Many Arguments and Experiments'), published in London in 1600. It is divided into six books, all dealing with roughly one subject but frequently digressing into seemingly tangential material. The first book examines magnetic substances. The next four books deal with the various known motions of lodestones and especially of the compass needle: coition, direction, variation and declination. In Book 6, Gilbert argues that the whole body of the earth is a magnet which revolves daily on its axis. This work, with its dismissal of Peripatetic philosophy and emphasis on an experimental method, led to Gilbert's reputation as the father of magnetism and as part of the vanguard of modern science.[1]

More recent historians have qualified these sentiments by exploring some of the less modern elements in Gilbert's philosophy (Freudenthal 1983; Henry 2001; Pumfrey 2002). Often referred to simply as *De magnete*, the full title of Gilbert's ground-breaking publication is a reminder that, beyond the investigation of magnetic motions and their causes, there is a larger cosmological agenda at work. This agenda reveals itself more fully in his other work, *De mundo nostro sublunari philosophia nova* ('A New Philosophy Regarding Our Sublunary World'), left unfinished at his death and known to have circulated in manuscript form before being published in Amsterdam in 1651.[2] *De mundo* seems to consist of the drafts of two distinct books: *Physiologia nova contra Aristotelem* ('A New Natural Philosophy against Aristotle') and *Meteorologia nova contra Aristotelem* ('A New Meteorology against Aristotle').[3] The *Meteorologia* is thought to have been started in the 1580s, with the *Physiologia* being composed later in the 1590s, although both were probably being revised by Gilbert up until his death.

To the historian who comes to know Gilbert first through *De magnete*, the *De mundo* seems somewhat of an anomaly, containing very little of the experimental method and mathematical sophistication of his more famous work. However, there

[1] See, for example, Whewell 1857, III, 37. More recently the comment by Kargon 1966, 9: 'It is, by now, a cliché to hail Gilbert as one of the first experimental scientists, but there is much truth to this well-worn example.'

[2] A manuscript version survives in the British Library, Royal MS 12 F XI.

[3] The manuscript version only gives a title for the *Meteorologia*, and so the title *Physiologia* may have been an addition by the publisher.

are certain issues pertaining to the composition of *De magnete* which shed some light on this seeming imbalance. Stephen Pumfrey argues that *De magnete* was somewhat of a collaborative work, with some of the more technical and mathematical features being contributed by some of the colleagues whom Gilbert entertained at Wingfield House, such as Edward Wright (1561–1615).[4] From Pumfrey's perspective, it is *De magnete* that is somewhat of an anomaly, with *De mundo* being a more accurate representation of Gilbert's own philosophical style. No conclusion can be definitive considering the unfinished nature of *De mundo* and the lack of source material that might shed light on the composition of *De magnete*. What seems apparent, though, is that, considered together, Gilbert's three works, *De magnete*, *Physiologia nova* and *Meteorologia nova*, represent a single project. That project was the establishment of a new cosmology and natural philosophy to supplant those of Aristotle.

Among Gilbert's main doctrinal targets in this mission against Peripatetic philosophy were the four terrestrial elements, the impossibility of a void, the immobility of the earth, and the distinction between the celestial and terrestrial realms. In Gilbert's cosmology, 'true earth', with its nascent magnetic potency, is the only true element in the sublunary world. The earth, planets and stars (which Gilbert called collectively the primary globes of the world) each have their own unique nature but are in many senses all of a kind. All are surrounded by their effluvia, some (such as the sun and the brighter stars) produce light, and they all transmit their natures outwardly into an 'orb of virtue' (*orbis virtutis*). In between the globes is the vacuum, which for Gilbert was complete privation and offered no resistance. The globes move freely through this void, their motions directed by a soul with the aim of regulating celestial influences in order to conserve, perfect and ornament their bodies. Gilbert's understanding of celestial motions, including those of the earth, was built upon two consequences ensuing from his decision to abandon the distinction between the earth and the heavens. These were a radical reinterpretation of astrological physics and the assumption that the surfaces of all the globes were adorned in a way similar to the earth.

3.2 The Sublunary World: Elements, Magnetism, Electrical Attraction and the Soul of the Earth

Considering that Gilbert's theories about the celestial world were largely built upon analogies with terrestrial physics, a proper analysis of those theories (including his stance on celestial influence and inhabitation) must begin with an examination of his philosophy of the sublunary world. In the opening chapters of *De mundo*, Gilbert dismissed the idea that everything below the orb of the moon was composed of the

[4]The main evidence comes from Mark Ridley, an associate of Gilbert's who worked on magnetic philosophy. He tells us that Edward Wright authored at least one chapter of *De magnete*, and one 'Doctor Gissope'—most likely Joseph Jessop (c. 1561–1599)—also assisted Gilbert, whose own knowledge of Copernicus was limited. See Ridley 1617, 9.

four elements. These elements, he argued, have never been seen, and cannot be demonstrated to exist in the form that Aristotle imagined. The idea that each element is a combination of two of the four qualities (hot, cold, dry, moist) was also fanciful and counter-intuitive (Gilbert 1651, 10–12). Rather, Gilbert maintained that the only substance worthy of being called an 'element' in the traditional sense was 'true earth'. This is the homogeneous matter that makes up the majority of the solid mass of the globe of the earth. All the other substances seen towards the periphery of the globe are created by the degradation of this 'true earth'. Both water and air are thus considered as exhalations or effluvia of the earth. Gilbert had explained this in *De magnete*:

> The earth emits various juices, which are not produced from water, or dry earth, nor from their mixtures, but rather born from the substance of the earth; these juices are not distinguished by contrary qualities or substances, nor is the earthly substance simple, as the Peripatetics dream. The juices come from vapours, risen up from great depths; indeed, all waters are extractions or exudations, so to speak, of the earth.[5]

The difference between earth, water and air, therefore, is one of 'state', in something approaching the modern sense of the term, rather than nature or quality in the Aristotelian sense. Earth can break down, attenuate or expand into other substances such as water and air, but, while this may be a process of degradation, the hypostatic principle of true earth remains within the substance (Gilbert 1651, 109). As for fire, this is described in *De mundo* as 'nothing other than intense heat, which has entered into a suitable material; an activity inherent to the perishing humour. Therefore it is not an element, nor a principle of nature'.[6]

There are several other factors crucial to Gilbert's understanding of matter. He rejected the possibility of a vacuum within substances and was strongly opposed to atomistic philosophy. This position he makes clear in *De mundo*:

> Body adheres to body, not due to the abhorrence of a vacuum. Bodies are separated by bodies, as solid bodies on earth are by the air and water flowing through them. This happens so that the continuity of bodies may be preserved, and therefore they do not admit a vacuum in their innermost parts.[7]

And again:

> We know that denser bodies change into more tenuous ones, as water extends into air with heat... Yet neither water nor air are composed by atoms or the aggregation of bodies, but are continuous bodies, and united in all their parts.[8]

---

[5] Gilbert 1600, 20: 'Terra emittit succos varios, non genitos ex aqua, aut terra sicca, nec ex earum mixturis, quam ex telluris substantia prognatos, hij non adversis qualitatibus, aut substantiis distinguuntur, neque tellus substantia est simplex, ut somniant Peripatetici. Existunt succi ex sublimatis ex profundioribus locis vaporibus. Aquae etiam omnes, telluris sunt extractiones et quasi exsudationes.'

[6] Gilbert 1651, 20: 'nihil aliud quam intensus calor, qui materiam idoneam invasit, actus inhaerens perituro humori. Non igitur elementum, nec naturae principium'.

[7] Gilbert 1651, 53: 'Corpori adhaeret corpus, non propter fugam vacui, separantur corpora per corpora, ut apud nos solida, aere aut aqua interfluente: quod sit ad continuitatis conservationem, atque ita vacuum in penetralibus non admittunt.'

[8] Gilbert 1651, 65: 'Sed nos cognoscimus crassiora corpora in tenuiora transire, ut aquam per calorem in aerem extendi... Neque enim aqua, aut aer, ex atomis aut corporum aggregatione constant, sed sunt corpora continuata, et omnibus suis partibus unita.'

In Gilbert's view, it is humour or moisture that is responsible for maintaining cohesion in solid bodies (Gilbert 1651, 104). At a later part of *De mundo*, he listed what he considered to be qualities: heat, moisture and dryness. Cold, rather than being a quality, is simply the privation of heat. Yet even for the three remaining qualities, Gilbert was keen to curtail their active ability. On moisture and dryness, he stated: 'They are not things themselves or primary operating qualities. Bodies are not changed by moisture or dryness, but by the application or detraction of moist or dry bodies.'[9] So if these qualities are not responsible for the alteration of bodies, what is? Gilbert listed these things: 'Essences, activity, and the privation of activity change bodies; not qualities.'[10]

The Aristotelian system of four elements and their composite qualities was therefore reduced to a monist philosophy, with the qualitative distinction between earth, water, air and fire replaced by a single spectrum of rarity and heat based on the homogeneous substance of the globe of the earth. This homogeneous substance is endowed in Gilbert's philosophy with two main attributes. One is an intrinsic moisture, variously labelled *succus* ('juice' or 'sap'), *humor* or *fluor* ('fluid substance'). It is the attenuation of this humour which results in the effluvia of water and air, with heat produced at the same time. Indeed, this was Gilbert's very definition of heat:

> We define all heat as the activity of a body from the attrition of humour, either intrinsic or external, for it is evident that no body grows warm without the attrition of humour, nor is humour abraded without heat being immediately produced.[11]

The other attribute was what he described in *De magnete* as the primordial and 'prepotent' form of the earth: magnetism (Gilbert 1600, 42). One of the main achievements attributed to Gilbert in the history of science is his assertion that the entire body of the earth is a dipole magnet. He believed that magnetism was a force intrinsic to true earth, the pure substance forming the majority of the globe. 'This globe of ours', he stated in *De mundo*, 'which consists of earth and water along with the other effluvia, is imbued with magnetic virtues, principally from its solid and firmer primary part.'[12] In Gilbert's understanding, this virtue is manifest in those

---

[9] Gilbert 1651, 216: 'non sunt res ipsae aut primae qualitates operantes; nec immutantur corpora per humiditatem et siccitatem, sed per appositionem et detractionem corporum humidorum aut siccorum'.

[10] Gilbert 1651, 216: 'Essentiae, et actus, et actus privatio, mutant corpora, non qualitates.' Another possible translation is 'Both the activity of an essence and the privation of that activity change bodies; qualities do not.' The concept of *actus* is central to Gilbert's philosophy, but it is difficult to find an English term that appropriately conforms to his varied usage. In Aristotelian philosophy *actus* is the Latin translation of ἐνέργεια (*energeia*), and is often translated into English as 'actuality'. In Gilbert's philosophy, the meaning more often approaches that of Aristotle's 'second actuality' (*On the Soul*, II.1, 412a22), comparable to the Latin *operatio*. As a compromise, it is translated as 'activity'.

[11] Gilbert 1651, 78: 'Quare calor omnis a nobis definitur, ut sit actus corporis ab attritione humoris insiti vel alieni. Nullum enim incalescere corpus sine humoris attritione manifestum est, nec atteritur humor, qui illico non sit calidus.'

[12] Gilbert 1651, 35: 'Globus hic noster, qui ex terra et aqua simul cum effluviis aliis constat, praecipue ex solida et firmiori parte primaria, magneticis imbutus est viribus.'

substances which are the nearest to that pure earth, such as the lodestone or iron. He consequently considered iron to be far more venerable than either gold or silver (Gilbert 1600, 24).

De magnete is structured to explain various known magnetic motions of lodestones and compass needles which had navigational, and to Gilbert also geological and cosmological, implications. These five motions are coition (here he rejected the more common term 'attraction'), direction, variation, declination and revolution. Gilbert's aim was to show that they all derived from a 'congregant nature'. Magnetic motions, he argued, 'are the incitation of homogeneous parts either towards one another or towards the primary conformation of the whole earth'.[13] Earlier philosophers had argued that the compass needle was directed either to some point in the heavens, such as the north Pole star, or to some large magnetic mountain or continent in the far north of the globe. Gilbert instead insisted that it is the body of the earth itself that has a magnetic polarity responsible for the movement of a compass, and, significantly, that it projects out from itself a magnetic orb of influence.

It is this magnetic virtue of the earth, as Gilbert explained in the last two books of *De magnete*, that enables the earth to rotate diurnally while maintaining a fixed verticity. True earth is not inert or passive as it was seen to be in the Peripatetic philosophy; rather it is an inherently active and mobile substance:

> We consider true earth to be a solid substance, homogeneous to the earth, firmly bound together, and endowed (as in the other globes of the world) with a powerful primary form. It maintains its orientation with a fixed verticity, and revolves with a necessary motion caused by an inherent tendency to turn.[14]

While magnetism is a property of the solid matter of true earth, it is in itself a completely immaterial force. Unlike Descartes, who later hypothesised the existence of screw-shaped magnetic particles, Gilbert maintained that magnetism was, in the early modern understanding of the term, an occult virtue.[15]

Yet this magnetic force was for Gilbert more than *just* an occult virtue; it pointed to something more profound again. The motive ability of a magnetic substance suggested a form of animism, present not only in the lodestone or iron but in the entire substance of the earth itself. In *De magnete* he raised this idea in Book 2, Chapter 4, 'On Magnetic Force and Form, What It Is, and on the Cause of Coition'. He discussed the manner in which iron, when heated, loses its magnetic qualities, yet when it cools it regains a polarity in line with either that of the earth or an adjacent lodestone. To Gilbert this was clearly not a corporeal action, but a disposal and conformity due to primary form. It was not violent, but harmonious and concordant, and here he saw an analogy with the celestial spheres whose motion and disposition

[13] Gilbert 1600, 45: 'motiones magneticae... incitationes sunt partium homogenearum aut inter se aut ad totius telluris conformationem primariam'.

[14] Gilbert 1600, 42: 'Sed terram veram volumus esse substantiam solidam, telluri homogeneam, firmiter cohaerentem, primaria, et (ut in globis aliis mundi) valida forma praeditam; qua positione, certa verticitate constat, et insita volubilitate motu necessario volvitur.'

[15] On Descartes' theory of magnetism, see his *Principles of Philosophy* (1644), IV, articles 133–83; see also Gaukroger 2002, 173–79. On the change in the meaning of the term 'occult', see Hutchison 1982.

are similarly regulated. 'For the power to move oneself seems to indicate a soul,' he reasoned, 'and the supernal bodies, which are also celestial and, as it were, divine, are thought by some to be animate, because they move with an admirable order.'[16]

Like the motion of the planets and stars, which will be discussed further below, Gilbert considered magnetic motion to be indicative of an animate nature. Since he also believed that the rotation of the earth was magnetic, Gilbert saw fit to declare that the earth was in fact a living creature. He compared its physical features and motion to the organs and locomotion of animals, and drew a link between its own life-force and that of the things which live upon it:

> The earth has its own form, it has internal spirits which flow in and out, and the mind moves the mass (*mensque agitat molem*) … The soul is the activity of an organic body. In the earth not only is the activity internal through the whole mass, but also the vital forces of those things which are begotten, grow and exist on the earth depend upon it. If you need organs, look at the poles, the equator and the meridian, and the agreement of its homogeneous parts. It is possible to see veins of ores in its outermost parts, extraordinary repositories of salts, metals and stones. Furthermore, you can see the processes of things and orders of fibres in the deepest sections, especially in the homogeneous and magnetic parts.[17]

Gilbert argued that the observable characteristics of the substance of the earth—its primary magnetic form, its motion, and its 'breath' (the cyclical flow of water and air from the interior)—demonstrate that the earth itself has a soul. This animate nature is not confined to the centre nor any other point, but is present in all matter that has true earth as its informative principle.

Another force that Gilbert investigated in *De magnete* was electrical attraction. The most common form of this phenomenon discussed at the time was the attraction of chaff to rubbed amber. For Gilbert, a key distinction between the two apparent attractions of lodestone and amber was that the amber did in fact act materially through the surrounding air. He pointed out that magnetism still acted through intervening substances, while electrical attraction could be blocked by a thin cloth. The amber, when rubbed, emits a very fine effluvium, which draws other objects to it within a certain radius:

> For electrical effluvia have the power of, and are analogous with, extenuated humour; but they will produce their effect, that is, both union and continuity, not by the external activity

[16]Gilbert 1600, 68: 'Vis enim movendi sese anima ostendere videtur, corporaque superna quae et caelestia tanquam divina, censentur a quibusdam animata, quod ordine admirabili moveantur.'

[17]Gilbert 1651, 125–26: 'Habet etiam tellus suam formam, habet spiritus internos, qui effluunt redeuntque, mensque agitat molem… Anima est actus corporis organici. In tellure non solum actus est insitus per universam molem; sed etiam adnascentium, excrescentium, supervenientium entelechiae ab illo pendent. Organa si desideres, ecce tibi polos et aequinoctialem et meridionalem, convenientiam partium eius homogenicarum. Videre etiam licet in eminentioribus partibus venas glebarum, salium, metallorum, lapidum admirabiles thesauros, in profundissimis etiam et maxime homogenicis et magneticis fibrarum ordines et processus rerum.' *Mensque agitat molem* is a quotation from Virgil, *Aeneid*, VI, 727.

of humours, nor by the heat and attenuation of heated bodies, but by their moisture itself attenuated into its own peculiar effluvia.[18]

While the distinction between magnetic and non-magnetic bodies rested on their material proximity to true earth, the determining factor for electrical substances was moisture. The crust of the earth is made of both humid and dry matter. Those things which have 'originated predominantly from moisture' produce electrical attraction when rubbed; those made mainly from dry earth do not.[19]

Gilbert also illustrated another important distinction between the two forces: 'In all bodies in the world two causes or principles have been laid down, from which the bodies themselves are produced: matter and form. Electrical motions acquire their strength from matter, but magnetic motions from the primary form.'[20] These two disparate forces both play their part in the preservation of the globe of the earth, which 'is congregated and coheres throughout itself electrically' while being 'directed and turned magnetically'. At the same time, it is 'cemented in its innermost parts so that it may cohere and be solid'.[21] Just as he had done with magnetism, Gilbert took his small-scale experiments of electrical attraction and expanded them into broader cosmological explanations. Gravity, therefore, is explained by Gilbert in *De magnete* as an electrical attraction by the earth's effluvia: 'Air (the common effluvium of the earth) not only unites the disjointed parts, but also the earth calls bodies back to itself by means of the intervening air.'[22]

This was Gilbert's view of the earth, as described in *De magnete* and *De mundo*: a living entity, composed of one element, the 'true earth' forming the interior mass, its inherent moisture expanding into the effluvia of water and air, cohering internally through that same moisture, externally by the electrical attraction of the air, and being imbued with a magnetic potency which moves and directs it. But there were many more diverse phenomena to be explained, and it was the ambition of *De mundo* to explain as many of them as possible. A crucial aspect of his natural philosophy was one shared by the Peripatetics and Renaissance philosophy in general: celestial influence.

[18] Gilbert 1600, 59: 'Humoris enim extenuati vim et analogiam habent electrica effluvia, nec ab actu humorum externo, calore, et attenuatione calidorum corporum, sed per ipsum humidum attenuatum, in sua et peculiaria effluvia, effectum dabunt, et unitionem et continuitatem.'

[19] Gilbert 1600, 52: 'Omnia igitur quae a praedominanti humido orta sunt, et firmiter sunt concreta, et fluoris speciem, et naturam inclytam retinent, in corpore firmo et concreto: alliciunt corpora omnia, sive humida, sive sicca.'

[20] Gilbert 1600, 52–53: 'In omnibus mundi corporibus duae propositae sunt causae, sive principia, ex quibus ipsa corpora producta sunt, materia et forma; Electricae motiones a materia, magneticae vero a forma praecipua invalescunt.'

[21] Gilbert 1600, 60: 'Globus telluris per se electrice congregatur et cohaeret. Globus telluris magnetice dirigitur et convertitur; simul etiam et cohaeret, et solidus ut sit, in intimis ferruminatur.'

[22] Gilbert 1600, 57: 'Aer (commune effluvium telluris) et partes disiunctas unit, et tellus mediante aere ad se revocat corpora.'

3.3 The Superlunary World: Void and Celestial Influence

In *De mundo*, Gilbert made clear that much of the activity on the surface of the earth is brought about through the action of the celestial bodies, primarily the sun and the moon. It is the light of the sun which warms the earth, attenuating the earthly humour into water and air: 'The sun, due to various motions in nature, either by diurnal revolution or the annual course, illuminates all things in all regions. It brings forth heat by light, and elevates air throughout the whole orb of the earth.'[23] In fact, in Gilbert's new philosophy the various influences of the celestial realm are largely responsible for the deterioration of the substance of true earth and its transformation into the variety of matter we see on the periphery of the globe. But investigating Gilbert's theories on celestial influence first requires a study of one of his major points of departure from Aristotelian cosmology: the possibility of a vacuum.

While Gilbert explicitly rejected the existence of a vacuum *within* bodies, in *De mundo* he was adamant that above the air, which extends only a few miles above the earth, there is the void. He first argued this on the basis that the space between the earth and the sphere of the moon cannot possibly be taken up by the elements: 'Therefore, the world is not filled in that whole space by a sufficient mass of elements, nor do the heavens descend below the orb of the moon. There is therefore a vacuum.'[24] He also argued that, if the heavens were filled with some sort of substance, like ether, then it would interfere with the passage of light from the far distant stars, no matter how fine or diaphanous it was: 'No body is so transparent that there is no resistance in it. Moreover, all resistance is an impediment and causes light to refract.'[25] Gilbert used these arguments and several similar to establish, contrary to accepted doctrine, that there is indeed a vacuum separating the celestial globes. In *De magnete*, this topic was only briefly mentioned when he talked of motion in 'an ether which is free of resistance, or else in a vacuum.'[26] In *De mundo* he also gave the reader a choice of term on occasion, but, when making his arguments, he was explicit in his terminology: 'est igitur vacuum'. Furthermore, his definition of the vacuum was clear and simple. It involves the absolute privation of bodies and activity (Gilbert 1651, 50).

So nothing exists in the vacuum, yet the vacuum allows the light from the sun and the stars to reach the earth. The nature of light itself was another subject of interest to Gilbert. He made a distinction between *lux* and *lumen* similar to that made by Patrizi. The two terms are often translated as 'light' and 'brightness' respectively, but perhaps an easier translation to work with in Gilbert's case is to

---

[23] Gilbert 1651, 29: 'Sol propter varios in natura motus, undique aut diurna revolutione, aut annuo cursu omnia lustrat, caloremque lumine accendit, aereamque elevat materiam per omnem orbem terrarum.'

[24] Gilbert 1651, 45: 'Non igitur in illo spatio toto plenus est mundus elementorum sufficienti mole, nec coelum infra orbem Lunae descendit. est igitur vacuum.'

[25] Gilbert 1651, 52: 'Nullum corpus ita diaphanum est, quin aliqua sit in illo renitentia, omnis autem renitentia impedimentum est et luminis refractio.'

[26] Gilbert 1600, 221: 'in aethere a renitentia libero, aut in vacuo…'

think of *lux* as fire or flame and *lumen* as light.[27] This was one of his explanations in *De mundo*:

> Light (*lumen*) is an existing thing. Flame (*lux*) is an activity rising from attenuating humour. Light is an activity proceeding from flame. Light is an immaterial being, effused by flame, received by a body. Light crosses the ether, but does not stay or exist in it. It is restrained by dense and opaque bodies.[28]

Light travels through the vacuum instantly, *sine tempore*, until it slows down or is stopped by contact with matter. Due to this instantaneous motion, it cannot be said to really exist in the vacuum, having no presence or effect until retained by an object. Gilbert argued, for example, that 'light, which spreads from the fire of the sun, is not warm in itself until it strikes a body; the sense of heat comes from the attenuation of the internal or near surrounding humour'.[29]

The light from the sun and the other stars played a very important role in Gilbert's theories of generation and corruption. He needed to explain how his essentially monist philosophy could account for difference and change. This is an issue he tackled directly in *De mundo*:

> The body of the earth is one and uniform. However, it is not simple in its qualities (as the opinion of many conceives them). A great variety of things blossom on its surface, stirred up both by its internal forces and by the lights of the heavenly bodies, and it suffers some imperfection and blemishing of its purity to a depth of several miles so that its upper regions may flourish, admirably adorned with new forms of things. The primary substance of the earth, however, is firm, united and stable, the whole of which is magnetic and filled throughout. It is not arid and sapless, like chalk, or filled with inert ashes, but it has its own sap and intrinsic humour, and in this way all things on its surrounds are born or similarly generated. A greater quantity of fluid matter is evident on the exterior than in the interior, because in the interior the humour is attenuated by activity into spirit and tends towards the surface, where it is turned into water, as if the waters were the sweat and effluvia of the mass of the earth. But it is by the lights of the stars, especially the sun, that the humour is continuously elaborated and driven through the whole circumference of the earth. And although the humour seems to us to be extinguished by the heat of the sun, it is in fact attenuated, and once again will return to its pristine form, and fall back down in another region of the circumference.[30]

---

[27] Gilbert 1651, 214: 'Lux est eadem quae flamma.' Compare this with Patrizi 1591, fol. 101^r: 'Lux enim flamma est. Et flamma est lux.'

[28] Gilbert 1651, 214: 'Lumen est ens. Lux est actus assurgens ab humore attenuante. Lumen est actus procedens a lucido. Lumen est ens immateriale, effusum a luce, a corpore exceptum. Lumen transit aetherem, nec in eo manet, aut est, a denso et opaco sistitur.' The verb *sisto* could have several meanings in this context. It could mean that the light is halted by opaque bodies, but it could also mean it is sustained, made to appear, or conveyed to our senses by them. This theory of light places Gilbert in the Platonic tradition of Plotinus, Ficino and Patrizi, as opposed to the scholastic tradition which considered light as a quality of the medium. See Lindberg 1986.

[29] Gilbert 1651, 89: 'Lumen, quod a Solis luce dimanat, per se calidum non est, donec corpus feriat, humorem insitum aut proxime circumfusum atterendo; unde caloris ille sensus.'

[30] Gilbert 1651, 108: 'unum enim telluris est corpus et uniforme, non tamen qualitatibus, ut multorum eas opinio concipit, simplex: etiamsi et insitis viribus, et luminibus astrorum concussum in eminentiis, multa rerum varietate efflorescat, et ad milliariorum quorundam profunditatem, in sua integritate, ut novis rerum formis superna admirabili ornatu floreant, labem aliquam et imperfec-

The light from the sun and the stars is responsible for the attenuation of the humours of the earth and therefore a large amount of the activity and variety on the surface of the earth. Light, however, is not the only thing which can stretch across the vacuum, nor is it the only form of celestial influence. The key to understanding this other kind of influence is once again magnetism. In *De magnete*, Gilbert gave an explanation of the transmission of the magnetic force that was very similar to that of light:

> The magnetic virtue flows out in every direction around a magnetic body in an orb; spherically around a terrella, more confusedly and unevenly in other shapes of stones. Yet there exists in nature no orb, or permanent or essential virtue spread through the air, rather a magnet only excites magnetic substances that are at a convenient distance. And as light arrives in an instant (as the experts on optics teach), so even more so is the magnetic vigour present within the limits of its power, and because its activity is much more subtle than light, and does not consent with a non-magnetic substance, it has no intercourse with air, water, or any non-magnetic substance; nor does it drive a magnetic substance to any motion by inrushing forces, but being present in an instant, it invites friendly bodies.[31]

In *De mundo* Gilbert categorised magnetism as the 'effused form' of the earth. Effused forms, similarly to light, traverse the void but do not stay or exist in it. But unlike light, which is stopped by a dense and opaque body (*a denso et opaco sistitur*), an effused form is not stopped by anything (*a nullo resisti potest*) (Gilbert 1651, 214–15). Gilbert used this theory of effused forms to explain, for example, how the moon affects the tides:

> The astral virtue of the moon extends beyond the limits of its body. The seas are moved not by its own light or borrowed light, or by its motion, but by the activity of the astral virtue, while the earth is within the orb of its virtue.[32]

The effused form, which extends outwards from the moon and is not stopped by dense or opaque objects, explains why the moon still has an effect on the earth

tionem sentiat. Primaria autem telluris substantia, firma, unita, et stabilis est, quae tota magnetica, interiora implet, non arida et exsucca, ut calx, aut cinis coacervatur iners; sed succum habet suum et humidum insitum, quaemadmodum ea. omnia, quae in ambitu eius aut crescunt, aut quovis modo generantur. Vis tamen materiae fluidae, in eminentioribus magis apparet quam intimis, quod in internis actus humor attenuatus in spiritum et eminentias tendit, ubi in aquam vertitur, quasi aquae terrenae molis sudores essent et effluvia. Sed astrorum lumine, vel maxime Solis, per ambitum terrarum humor elaboratur et urgetur continuo; qui etsi nobis evanescere aliquando propter Solis ardores videatur, attenuatur tantum, et iterum in pristinam naturam revertitur, et in aliquam circumferentiae regionem relabitur'.

[31] Gilbert 1600, 76–77: 'Funditur virtus magnetica undequaque circa corpus magneticum in orbem; circa terrellam sphaerice; in aliis lapidum figuris, magis confuse et inaequaliter. Nec tamen in rerum natura subsistit orbis, aut virtus per aerem fusa permanens, aut essentialis; sed magnes tantum excitat magnetica convenienti intervallo distantia. Atque ut lumen in instanti advenit (ut docent optici); ita multo magis vigor magneticus intra virium terminos praesens est, et quia eius actus multo quam lumen est subtilior, et cum non magnetico non consentit, cum aere, aqua, aut quovis corpore non magnetico nullum habet commercium, nec magneticum commovet motu aliquo irrumpentibus viribus, sed praesens in instante amica corpora invitat.' 'Terrella' is the name given to a spherical magnet.

[32] Gilbert 1651, 186: 'Astrica virtus Lunaris ultra corporis limites extenditur. Non enim lumine aut proprio, aut mutuato, aut motu, maria commovet, sed actu astricae virtutis, dum terra intra orbem virtutis sit.'

throughout all its phases, or when it is below the horizon. Every celestial body, just like the lodestone or the magnetic earth, effuses its form outwards into an orb of influence or virtue. Gilbert differentiated the effect of these effused forms from that of light, whose main purpose is to attenuate humour and create heat:

> Warmth does not arrive from above except through light. In the earth, light attenuates the earth's humours and excites heat; the various lights from the stars have therefore various effects. The other astral virtues, however, are not produced by heat, because the heat arising from light is of the same nature and has the same effect as other heats. Humour is stirred up by light in the same way as it is by heats engendered here on earth ... It should be known that there is an effect of a different nature, deduced from form, not material, nor arising from heat or light. There are primary forms in the stars, singular and specific; as in the earth there is the powerful and excellent magnetism. By this cause the moon, when it is new and without light, still affects the earth and draws humour... Likewise, by this cause the planets, which shine both by their own light and the light of the sun, affect the earth in many diverse ways; Venus different from Saturn, Jupiter and Mars different from Mercury or Venus. For if they were to be without light, like the moon, the effect would still be made. Likewise, and in the same way, the fixed stars emit certain influences, which are like the seedbeds of many things, which cannot be blocked by any impediment, no different from the way that the primary magnetic virtue of the earth and its parts have their activity unobstructed. It will be helpful to know the activities of the stars, which are not less illustrious than this extraordinary power, attested to by stunning forces (magnetism), but are distinguished by their great and primary properties. Therefore, always different natures descend from the stars: from Lyra different from Arcturus, from Hyades different from Canis Maior, from Praesepe different from Spica in Virgo, or from the Cor Leonis.[33]

Gilbert's theories concerning the efficacy of celestial influence resemble the 'multiplication of species' of Grosseteste and Roger Bacon. Yet they differed radically from traditional scholastic astrology by holding up effused forms (based on the example of the magnet) as a secondary form of influence *separate* from light. Unfortunately, Gilbert's astrological physics was not developed in *De mundo* in a particularly systematic or coherent way. Many of the most crucial concepts are found at the very end of the *Physiologia* amongst somewhat miscellaneous collections of aphoristic notes. For example, Chapter 27 entitled 'De effluviis' ('On

[33] Gilbert 1651, 80–81: 'tepor vero nullus desuper dimanat, nisi per lumen; Lumen vero in tellure humorem telluris attenuat, calorem excitat; unde ab astris lumina varia varios habent effectus. Virtutes vero astrorum aliae non a calore fiunt. calor enim adveniens ex lumine eiusdem est efficientiae et naturae cum caeteris caloribus, agiturque humor eodem modo per lumen, sicut per calores hic apud nos prognatos... sciendum est, quod illud alterius naturae est, a forma deductae, non materiatae, nec a lumine aut calore provenientis. Sunt enim in astris formae primariae, singulares et propriae; sicut in tellure, magnetica praepotens et egregia. Ob eamque causam Luna cum sine lumine fuerit in interluniis, tellurem concutit, et humorem ducit... Ob eamque etiam causam errones longe diversis modis tellurem et corpora afficiunt; Venus aliter quam Saturnus, Iupiter Marsque aliter quam Mercurius aut Venus; qui et propriis luminibus, et solaribus etiam fulgent. Quod si nulla haberent lumina, ut Luna, effectus tamen facerent. Eodem etiam modo stellae fixae influentias quasdam immittunt, tanquam rerum plurimarum seminaria, quae nullis obicibus refraenari possunt; non aliter atque primaria magnetica virtus telluris et eius partium, efficientiam habent a nullo impeditam. Ab hac tam egregia et demonstrata miris virtutibus potentia, astrorum actus non minus illustres, sed summis et primariis dotibus insignes agnoscere juvabit. Quare alia atque alia pendet ab astris natura; a Lyra alia quam ab Arcturo: ab Hyadibus diversa a cane majori: a Praesepi aliena a Virginis Spica, a corde Leonis...'

Effluvia') begins: 'In the effluvia of the sun and the moon, the rays of the fixed stars mix, and descend to the earth with the light of the sun, and the borrowed light of the moon.'[34] On the surface at least, this idea is similar to those found in Cusa and Patrizi, where the sun and moon serve as intermediaries for the influence of further-flung bodies (see Sects. 2.3 and 2.5), but Gilbert provided no further explication. One page later, on *lux* and *lumen*, Gilbert stated: 'Light (*lumen*) penetrates every body, even the most opaque ones to a certain extent, by which intrusion it [i.e. the body] is deformed.'[35] Two pages further in a chapter on forms and qualities, he presented the aphorism that 'activity destroys the form of substance.'[36]

As cryptic as some of these statements may seem, a detailed reading of the whole work reveals at least the supporting structure of Gilbert's new understanding of celestial influence. The two kinds of conveyor of this influence are light and the effused forms, both of which cross the vacuum between the globes instantaneously, without presence or action, until interacting with a material body. Both are responsible, in their own way, for degrading and transforming the homogeneous substance of the globe of the earth into the variety of substances seen on its surface. The activity of light seems to be restricted to the attenuation of humour (and subsequently the production of heat), although Gilbert did allow for some diversity amongst the celestial *lumina* regarding this outcome. Effused forms, being unique and fitting to each celestial body, permit a much broader range of possible influences.

Gilbert's new 'physiology' of the sublunar world, by dispensing with the doctrines of the four elements and qualities, eroded much of the explanatory foundations of traditional astrology. Indeed, he occasionally criticised certain aspects of contemporary astrological practice in *De mundo*, but always remained laudatory of the science itself.[37] Early in the text, for example, he wrote: 'Astrologers, in a way that is not so much imprudent as insane, attribute the elementary qualities to their Trigons, and establish four of them for the four elements; and many of these trifles in Astrology, the queen of the sciences, are fictions from ignorant soothsayers.'[38] Astrology retained an important place in Gilbert's new cosmology, not only in the abstract sense of celestial influence, but also being pragmatically relevant to issues ranging from meteorology to personal characteristics and health, and to geography and politics.[39]

[34] Gilbert 1651, 213–14: 'In effluviis Solis et Lunae, fixarum stellarum radii miscentur, et in terras descendunt cum lumine Solis, et Lunae mutuato lumine.'

[35] Gilbert 1651, 214: 'Lumen ingreditur omne corpus vel opacissimum aliquantulum, quo ingressu foedatur.'

[36] Gilbert 1651, 216: 'Actus destruit formam substantiae.'

[37] On astrology in England during this period, see Curry 1989; Dunn 1994.

[38] Gilbert 1651, 6: 'Astrologi enim non tam imprudenter, quam insane, Trigonis suis elementarias qualitates attribuunt, quatuorque instituunt pro quatuor elementis; atque eius farinae plurima in Astrologia, scientiarum regina, ab indoctis ariolis ficta sunt.'

[39] Interestingly, the best evidence we have for placing Gilbert's date of birth is a nativity, identified among the Ashmolean manuscripts by S. P. Thompson. See the discussion in Roller 1959, 64. The nativity in question is Bodl. MS Ashmole 176, VI, 3.

The only evidence we have of Gilbert engaging personally and practically with astrology is a section of the *Meteorologia* claiming to be a sample from a notebook of observations he had made of the weather and planetary positions. This notebook may be from his early days at Cambridge and so it is impossible to say whether his novel theories of celestial influence had begun to form at this stage. Pumfrey argues that the notebook represents Gilbert's disillusionment with Aristotelian meteorology in the early stages of his Cambridge education, but there is not enough information to know for certain (Pumfrey 2002, 18–19). Regardless of the exact import of the notebook, there is ample evidence in *De mundo*—in *Physiologia* as well as *Meteorologia*—that he still considered astrology relevant to meteorology. For example, he wrote:

> The winds are raised up by the earth itself, by the light of the sun, which impels most of all, by the light and specific forms of the stars and planets, and also by the moon, not so much through its borrowed light as by the effort and co-operation (*synentelechia*) of its unique virtue.[40]

Other examples of his regard for astrology abound within Gilbert's texts. One important passage in *De magnete* compared the effect of the magnetic earth upon iron to that of the heavens upon a new born:

> For the magnetic effluence of the earth rules everywhere within the orb of its virtue, and changes bodies; but it especially rules and controls those bodies which are by nature more similar and connected to it, such as magnets and iron. Wherefore in many affairs and actions it is clearly not superstitious and vain to observe the positions and conditions of regions, the points of the horizon and places of the stars. For as when a baby is brought forth from its mother's womb into the light, and acquires respiration and certain animal activities, then the planets and celestial bodies, according to their position in the universe, and according to that configuration which they have with regard to the horizon and the earth, instil peculiar and individual qualities into the new-born; so that piece of iron, whilst it is being formed and lengthened out, is affected by the common cause (i.e. the earth); and again when it returns from ignition to its former temperature, it is imbued with a specific verticity as a result of its position.[41]

Gilbert also employed astrology for explanations on a much larger scale. He used astrological reasoning to explain why the northern hemisphere is more suited to the activities of commerce, to the seats of republics, to the establishment of empires, and to the flourishing of the arts and disciplines (Gilbert 1651, 134).

[40] Gilbert 1651, 33: 'Excitantur venti a tellure ipsa, a Solis lumine, quod maxime impellit, a planetarum et astrorum luminibus et peculiaribus formis, a Lunae etiam non tam mutuato lumine, quam singularis virtutis conatu et synentelechia.'

[41] Gilbert 1600, 142: 'Regnat enim ubique intra orbem virtutis suae, telluris magnetica effluentia, et immutat corpora: Quae vero sunt illi natura magis similia et coniuncta maxime, regit et componit; ut magnetem et ferrum. Quare in plurimis negotiis et actionibus non est plane superstitiosum, et vanum, positiones et habitudines terrarum, horizontis puncta, et astrorum loca observare. Nam ut cum ex utero materno natus in lucem editur, et respirationem et animales quasdam actiones adipiscitur, tunc planetae et corpora caelestia pro habitudine sua in mundo, et pro ea. quam habent ad horizontem et terram configuratione, proprias et singulares nato immittunt qualitates: sic ferrum istud dum fingitur et extenditur, a communi causa (tellure scilicet) afficitur; dum etiam ab ignitione ad pristinam temperiem revertitur, verticitate singulari pro positionis ratione imbuitur.'

On an equally broad scale and time frame was Gilbert's astrological exposition of the phenomenon of precession. The precession of the equinoxes, the slow shift of the sun's equinoctial position against the fixed stars over thousands of years, was explained in Ptolemaic astronomy as a movement of the celestial sphere. First Copernicus and then Gilbert explained it more correctly as a slow change in the orientation of the earth's axis. In Gilbert's philosophy, there is a very clear reason for the earth's various motions, which was explained both in *De magnete* and *De mundo*. These motions, like that of the magnet, are for the earth's own benefit, so that the influences from the sun, moon, planets and stars do not become harmful, but are tempered and varied and make the earth fruitful (Gilbert 1651, 159). Thus he wrote in *De mundo*:

> The fixed stars also have their own powers, which work upon the earth. While admittedly they are weaker, over a long time… they could impart a more serious and disruptive change. For this reason, nature in its providence has another slow motion… in the inflexion of the axis of the earth.[42]

In *De magnete* Gilbert was more specific about the astrological implications of precession:

> Thus all the stars change their rays of light at the surface of the earth, through this wonderful magnetical inflection of the earth's axis. From this come new variations of the seasons of the year, and this is why regions become more fruitful or more barren; hence the characters and manners of nations are changed; kingdoms and laws are altered, in accordance with the virtue of the fixed stars, and the strength thence received or lost from the singular and specific nature of each as they culminate; or on account of new configurations with the planets in other places of the Zodiac; on account also of risings and settings, and of new concurrences at the meridian.[43]

An interesting parallel can be found in the *Narratio prima* of Georg Joachim Rheticus (1514–1574). In this first account of Copernicus' astronomy published in 1540, 3 years prior to *De revolutionibus*, the author Rheticus discussed the astrological relevance of another slow motion: apsidal precession. After he described Copernicus' account of the motion, he used it to offer an astrological interpretation of political history, and even gave an astrological prognostication, predicting the fall of the Mohammedan empire based on this slow migration of the sun's apogee.[44] For

[42] Gilbert 1651, 161: 'stellae fixae vires etiam habent suas, quibus tellus laboratur; quae licet minores sunt et exiles, tamen diuturnitate temporis, propter longiorem moram in parallelis circulis, unicuique perpendiculari mutationem magis gravem et molestam mitterent. ob eamque causam natura provida motum habet alium tardum, quia vis illata subitanea non est, axis telluris inflexione'.

[43] Gilbert 1600, 235: 'Sic omnes stellae immutant suos luminis radios in superficie telluris, admirabili hac magnetica axis telluris inflexione. Hinc temporum anni novae varietates, terraeque foecundiores magis[q]ue steriles evadunt; hinc gentium ingenia et mores immutantur, regna et leges alterantur, pro stellarum fixarum virtute, et robore suscepto aut amisso, pro singulari et specifica natura fixarum culminantium; aut propter novas in aliis Zodiaci locis cum planetis configurationes; propter ortus etiam, et occasus, et concursus in meridiano novos.'

[44] Georg Joachim Rheticus, 'Narratio prima', in Rosen 1939, 122. See also Westman 1975; Hooykaas 1984; Danielson 2006.

both Rheticus and Gilbert such motions needed more than physical and astronomical explanations, they needed functional and teleological justification as well. Celestial influence provided that. For Gilbert, magnetism showed 'how'; astrology showed 'why'.

Gilbert's interest in astrology has been paid very little attention in the few modern interpretations of his work. Pumfrey, in his biography of Gilbert, briefly mentioned that his 'profound belief in astrology was not unusual, but his belief that astrological forces were master-minded (mistress-minded?) by a terrestrial magnetic soul certainly was' (Pumfrey 2002, 170). As this section has attempted to demonstrate, there is much that was new and interesting about Gilbert's theories of celestial influence. Because he represents an early stage in the development of modern ideas of force, most discussions of his 'magnetic' philosophy, while recognising its debt to Platonic animism, fail to appreciate how this theory of the earth's astral virtue was part of a process of rethinking celestial influence. Gilbert's dismissal of three of the four elements, and of the four qualities as active forces, undermined the traditional understanding of astrology. This was a subject which, as a sixteenth-century natural philosopher and physician, we can assume he had a personal and professional interest in. At the same time, he gave a new dimension to the occult qualities of the celestial bodies by differentiating light from effused form in terms of both operation and effect.

The interposition of a vacuum between the globes did not limit the interaction between the celestial bodies. In fact, Gilbert saw it as essential, allowing the influences of each globe to arrive unimpeded and instantaneously at all the others. As he put it in *De mundo*: 'Because of the vacuum, things that are placed at seemingly infinite distances are essentially nearby. The immense universe acts everywhere in unison and harmony through all the globes, and things that may seem the most disconnected are in fact united, even to our limited human view.'[45] The notion that astral forces, as mentioned by Pumfrey, were mediated by the magnetic soul was indeed unusual, although not entirely unprecedented (see Sect. 2.2), and will be further discussed below. What truly separates Gilbert's astrology from the Aristotelian tradition, and aligns him more with Cusa and Bruno, is that the earth was no longer a passive recipient of astral forces, but a participant in the mutual exchange of influences between the primary globes of the universe.

3.4 The Void and the Motions of the Globes

Part of Gilbert's deconstruction of the edifice of Aristotelian natural philosophy was his criticism of the common understanding of natural motion. Natural motion, according to Aristotle, varied depending on the nature of the object in question (*On*

[45] Gilbert 1651, 73: 'Propter vacuum quae tanquam in infinita distantia posita sunt, prope adsunt: et immensus undique mundus, per omnes globos conspirat, concordat, disjunctissimaque uniuntur, vel cum mortalium parvis ocellis.'

the Heavens, I.2). For the incorruptible celestial bodies, composed of ether, this meant a perfect circular motion. Within the sublunary world, the elements moved rectilinearly in an attempt to find their place according to their disparate natures: heavier earth and water towards the centre; lighter air and fire towards the periphery. Gilbert fundamentally disagreed with Aristotelian notions of both place and motion, and proposed a different explanation for the tendencies of sublunary substances:

> Gravity is therefore the inclination of bodies to their own principle, such as the inclination towards the earth of those things which are cast from it, while levity is an incitation of a body by its own principle, either because the humour [inside] is thinning or because the surrounding bodies rise up.[46]

The principles of gravity and levity, however, only apply within the confines of a globe's effluvia, a point he made clear with the example of comets (Gilbert 1651, 49). When it came to discussing the movement of the globes themselves through the vacuum, Gilbert was even more dismissive of the importance of 'place'. The void was all important to Gilbert's new philosophy of the heavens. It contains no body, no activity, zero resistance. It is complete privation.[47] We have already seen its relevance to the question of celestial influence and the daily motion of the earth, which, with nothing to restrict it, revolves as one with its effluvia (Gilbert 1651, 142). The void also provides an open setting for the free motions of the other stars and planets:

> No sane person would say that the vacuum is the cause of motion, but in the separate vacuum bodies are moved; not by the vacuum, but by the bodies themselves, which emit their own activity. In the vacuum the prime bodies, the globes themselves, move, such as the planets and the earth, and even some of the fixed stars, if they have motion, which cannot be known due to their long distance from us.[48]

The vacuum cannot be a cause of motion itself. Nor, as Gilbert spent much time arguing, is there any such thing as a prime mover or any other solid celestial sphere carrying stars or planets. There is just the void, which itself is not surrounded by anything, but should be thought of as just continuous space (Gilbert 1651, 50). The planets and stars have a motive power, inherent throughout their homogeneous substance, as does the earth: 'The motive power (*motor*) of the earth is not in its centre more so than in its own or any other circumference (as a certain Nolanus thought), nor is the motive power in any other part.' The truth of the matter, Gilbert continued,

[46] Gilbert 1651, 47–48: 'Est igitur gravitas corporum inclinatio ad suum principium, a tellure quae egressa sunt ad tellurem. Levitas vero incitatio a suo principio, vel humoris solventis ratione, vel circumfusi corporis attollentis.'

[47] It is labelled as such in a diagram of the earth. See Gilbert 1651, 51: 'Vacuum separatum nullum Corpus, nullus actus, nulla renitentia, privatio.'

[48] Gilbert 1651, 48: 'Praeterea nemo sanus dixerit, vacuum esse causam motus, sed in vacuo separato moventur corpora, non a vacuo, sed ab ipsis corporibus, quae actus suos emittunt. In vacuo moventur prima corpora, globi ipsi, ut planetae, et terra, et fixae etiam aliquae stellae, si motum habuerint, qui propter longinquitatem a nobis discerni non possit.'

is that 'all homogeneous and primary parts, which are also magnetic, contribute to the motion'.[49]

Gilbert had already attempted to argue in *De magnete* that the magnetic power of the earth was responsible for its daily rotation. There were several reasons, however, why it could not be *entirely* responsible. Gilbert based many of his arguments about the earth on the reasoning that a small spherical magnet, or *terrella*, would behave in a similar way to the earth. The medieval study of magnets by Peter Peregrinus, a thirteenth-century natural philosopher, had done the same, although he had assumed that the behaviour mimicked the spherical heavens, not the earth. In his *Epistola de magnete* (1269), Peregrinus had argued that a terrella, suspended by its poles, would rotate once in 24 h. Such a result would have been valuable to Gilbert, but alas he could not recreate it:

> I omit what Peter Peregrinus constantly affirms, that a terrella suspended above its poles on a meridian moves circularly, making an entire revolution in 24 hours: which, however, has not happened to ourselves as yet to see; and we even doubt this motion on account of the weight of the stone itself, as well as because the whole Earth, as it is moved of itself, so also is it propelled by the other stars: and this does not happen in proportion (in the terrella) in every part.[50]

The second part of Gilbert's explanation is important for understanding his physics of planetary motion. Motion is achieved by the combination of the planet's own motive power with the propulsion provided by the other globes. The terrella does not replicate the earth's rotation because, being positioned on one part of the earth, it is not exposed to the same astral influences. In *De mundo* he expanded on the errors in Peregrinus' theory:

> Moreover, the earth is moved by the conspiring of the motions of the other bodies, and by their effused motive forms, especially those of the sun and the moon, by which the earth is contained and defined as if by the terms of a proportion… In this way Peter Peregrinus is mistaken.[51]

Apart from Peregrinus, there were other possible sources of inspiration for the application of magnetism to the question of the earth's motion. A contemporary of Copernicus, Celio Calcagnini (1479–1541), wrote an essay in about 1518–1519 with the title *Quod coelum stet, Terra autem moveatur* ('That the Heavens may

---

[49] Gilbert 1651, 165: 'Motor telluris non est in centro suo potius, quam in sua, aut aliena circumferentia (ut quidam putavit Nolanus) nec in ulla alia parte; sed omnes partes homogenicae et primariae, quae et magneticae, ad motum faciunt.' The *quidam Nolanus* is a reference to Giordano Bruno.

[50] Gilbert 1600, 223–24: 'Omitto quod Petrus Peregrinus constanter affirmat, terrellam super polos suos in meridiano suspensam, moveri circulariter integra volutatione 24 horis: Quod tamen nobis adhuc videre non contigit; de quo motu etiam dubitamus, propter lapidis ipsius pondus, tum quia tellus tota uti movetur a se, ita etiam ab aliis astris promovetur: quod proportionaliter in parte quavis (ut in terrella) non contingit.'

[51] Gilbert 1651, 138: 'Praeterea movetur terra conspiratione motuum aliorum corporum, et formis effusis commoventibus, praesertim Solis et Lunae, quibus proportionatur, et continetur quibusdam quasi terminis… Ita fallitur Petrus Peregrinus.'

stand still while the Earth moves').[52] In this essay Calcagnini directly compared the motion of the magnet to the daily rotation of the earth, and then linked this motion to terrestrial processes of generation:

> Do we not all admire the magnet stone, constantly turning one part to the south, the other to the north? ... From which is plainly understood how great the power of nature is, and how great its faculty that we call 'sympathy'. So then why does nobody admire the earth itself, which surely without the aid of the sun would lie inert and infertile, and so eagerly strives for the embrace of the sun, in order to obtain from it the vital flamelets for the propagation of things.[53]

Gilbert doesn't make any reference to this essay, but in *De magnete* he did refer to another essay by Calcagnini, *De re nautica*, which was published alongside *Quod coelum stet* in the *Opera aliquot* (1544) (Gilbert 1600, 7). This is not proof that Gilbert read it, as he may have had access to a different version of *De re nautica*, but there is a certain degree of probability. It was definitely not beyond Gilbert to avoid referencing other writers where it would detract from his own originality.

While there were precursors to his connection of terrestrial motion to magnetism, Gilbert takes the theory in new directions. Calcagnini, for example, had thought the earth more apt for motion than the heavens because the former was the place of change and mortality while the latter was nobler and more suited to rest (Calcagnini 1544, 389; Omodeo 2014, 211). Gilbert, on the other hand, continually stressed that the positions and motions of the globes have nothing to do with any predetermined notion of 'place', but are rather a product of forces between the planets themselves. So he wrote in *De mundo*:

> Place does not draw any bodies, rather the primary bodies distribute themselves through space by mutual activity... The maker of things... did not create one place around which things must remain or into which things must conglomerate, rather he gifted bodies with primary virtues, by which they mutually array themselves, and combine in an admirable order though the intervals of the universe.[54]

And again:

> [The stars] are not disposed due to place, or the power of place (for place is nothing), nor are they fixed to some more stable heaven, like nails to a vault, but they endure by the intrinsic forces between them... In the universe nothing wanders or is lost without a cogent and impelling cause.[55]

[52] The essay was only published after the author's death. *Quod coelum stet, Terra autem moveatur*, in Calcagnini 1544, 388–94. See also Omodeo 2014, 209–13.

[53] Calcagnini 1544, 390: 'Magnesium vero lapidem nonne omnes admiramur, parte una in austrum, in arctum alia pertinaciter inclinantem? Ex quo plane intelligitur quanta sit naturae vis, et eius quam *sympatheian* vocamus facultas. Ut iam non sit cur aliquis admiretur, terram ipsam, quae sane absque solis praesidio iners foret atque infoecunda, tam cupide in solis complexus eniti, ut ex eo vitales igniculos ad rerum propagationem mutuetur.'

[54] Gilbert 1651, 115: 'nec locus allicit corpora ulla, sed corpora ipsa primaria mutuis actibus disponunt sese per loca... Rerum igitur conditor... non singulis partibus primariis loca, circa quae, aut in quibus, conglobantur, et haerent, sed corpora ordinavit primariis virtutibus praedita, quibus mutuo disponunt sese, et per intervalla in mundo ordine mirabili combinantur'.

[55] Gilbert 1651, 112–13: 'Neque tamen ita repositae sunt propter loca aut loci vim, (locus enim nihil est) nec infixae caelo alicui firmiori, tanquam clavi in camerato, sed ab insitis permanent inter se viribus... In mundo enim nihil peregrinatur, aut vagum est, absque impellente et cogente causa.'

This understanding of planetary motion allowed Gilbert to abandon the epicycles and eccentrics of Renaissance astronomy. He, like many before him, took them to be convenient inventions by astronomers to help predict the movements of the planets. They were therefore of no use to the philosopher who looked toward causes. Skilled men like Ptolemy, Regiomontanus and Copernicus 'have their own systems and inventions of spheres' by which they save the appearances. But these are not real: 'The variety and plurality of the heavens was first of all invented by mathematicians, not philosophers.'[56] All perceived 'inequalities' in the motions of the celestial globes were for Gilbert the product of the varied and continually changing activities of the other globes. These motions, while not in perfect circles, are nonetheless certain and definite (Gilbert 1651, 187).

In Gilbert's cosmology, as we have seen, the heavens did not revolve around the earth in 24 h, rather the earth rotated on its own axis. The moon orbits the earth, but the planets, meaning Mercury, Venus, Mars, Jupiter and Saturn, do not orbit the earth, but rather the sun. Gilbert never offered an explicit opinion of whether the earth had an annual orbit around the sun or vice versa. In *De mundo* he repeatedly deferred the question, and used somewhat ambiguous language (Gilbert 1651, 49, 141). For the purposes of this discussion, the question of whether or not Gilbert was a Copernican is not as urgent as the 'how' and 'why' of celestial motion more generally.[57] The sun for Gilbert is certainly the most eminent among the globes in terms of its motive influence, but the nature of this power is not simply one of an active solar force working on passive planetary bodies. The motive virtues of the sun are in a sense *used* by the planets, which, by their own power, regulate their motion to their own advantage. This is the reasoning Gilbert saw behind both the rotation of the earth and the orbits of the planets:

> The earth is turned in the direction of the supernal bodies by a natural appetite for the preservation, perfection and ornamentation of its parts… The earth, by a certain great necessity and manifest virtue, intrinsic and conspicuous, is turned circularly to the sun; by which motion it enjoys the solar virtues and is firmed in its verticity, so that it does not revolve unstably to every region of the heavens. The sun is the most eminent actor in nature, as it drives the course of the planets, and here incites the revolution (*conversio*) of the earth. All these circular motions, both that of the earth and those of the planets, agree in one great cause, because the earth and the planets are all carried towards the east.[58]

---

[56] Gilbert 1651, 148: 'Ita viri peritissimi, Ptolemaeus, Regiomontanus, Copernicus, suas habent circulorum rationes et inventiones, quibus perpetua numerorum et apparentiarum consonantia correspondet. Non tamen circuli illi sunt corpora, sed spatia tantum imaginaria'; ibid.: 'Coelorum varietatem, et pluralitatem, primum omnium mathematici invenerunt, non philosophi.'

[57] On the subject of Gilbert's potential Copernicanism, see Koyré 1957, 284–86, n. 47; Freudenthal 1983, 33–35; Pumfrey 1987, 8–45; Gatti 1999, 98; Henry 2001, 106; Pumfrey 2002, 169–70. There remains no 'smoking gun' for Gilbert's commitment to an annual orbit of the earth. It is tempting (yet problematic) to make this conclusion for him based on his own physical principles. The allusions he makes in *De mundo* to astronomers past and present (including Copernicus and Tycho) are deceptive due to the fact that they are largely derived from similar passages in Patrizi. Compare Gilbert 1651, 149–51, with Patrizi 1591, fol. 91^{v}.

[58] Gilbert 1651, 142–43: 'Movetur naturali appetitu, ad suarum partium conservationem, perfectionem, et ornatum, versus praestantiora… Terra igitur magna quadam necessitate, et virtute manifesta, insita, et conspicua, convertitur ad Solem circulariter; quo motu solaribus virtutibus gaudet, firmaturque certa verticitate, ut non vage volvetur in omnem coeli regionem. Sol praecipuus in

He argued the same theory in the cosmological climax of *De magnete*:

> For since the sun himself is the agent and inciter of the moving universe, other wandering globes positioned within his powers, when acted on and stirred, also regulate each its own proper courses by its own powers; and they are turned about in periods corresponding to the extent of their greater revolution, and the differences of their effused powers, and their intelligence for higher good.[59]

The motions of the celestial globes, in terms of both orbit and rotation, are determined by their own intelligences. The sun provides the fuel, as it were, and the planets move in order to enjoy the virtues of the sun and the other globes in proper proportion.

John Henry has rightly pointed out that magnetism didn't work as an explanation of the earth's rotation without animism. It is this feature of Gilbert's cosmology which most readily invites comparisons with Cusa, Bruno and Patrizi. The pivotal argument of *De magnete*, which Gilbert sought to prove by experiments, is that the magnetism of the lodestone and hence of the earth is a product of an animate nature (Henry 2001, 117–19). In *De mundo*, Gilbert laid out his philosophy of the nature of celestial motion:

> In nature position is before movement. Therefore, God positioned the earth and the stars at certain distances, before they were put in motion. And this motion is from the impulse of the other bodies, and on account of the radiated virtues from these, as well as from the bodies themselves, when they seem to incline to betterment and perfection.[60]

All the globes have a soul, and in this sense Gilbert's astronomy was animistic, but the souls behave in a theoretically predictable way. Their motions are not haphazard or whimsical, but determined by their exposure to the radiated virtues from the other globes. The divine astral intelligence of the earth, no different to that of the other globes, directs itself so as to best enjoy these virtues, and thus ensure the 'preservation, perfection and ornamentation of its parts'.

---

natura actor, ut erronum promovet cursus, sic hanc telluris conversionem incitat; qui omnes motus in orbem, tam telluris, quam erronum, in una aliqua magna causa consentiunt, quod omnes, tam tellus, quam errones, versus ortum feruntur.'

[59] Gilbert 1600, 231: 'Nam cum sol ipse motivi mundi actor sit et incitator; globi alii errones intra vires eius positi, cum acti sunt et conciti, suis etiam viribus quisque cursus suos proprios moderatur, convertunturque suis temporibus pro vertiginis maioris amplitudine, et virium effusarum differentiis, et ad melius bonum intelligentia.'

[60] Gilbert 1651, 210: 'In rerum natura positio est ante motum. Ita Deus posuit tellurem, et astra, certis quibusdam distantiis, antequam moverentur. et hic motus sit impulsu aliorum corporum, virtutum ratione ab illius effluentium; vel a corporibus ipsis, cum ad melius, et ad perfectionem inclinare videntur.' Reading 'illis' for 'illius'. Compare this with a passsage from Cusa 1932, I, 112: '[Hunc opificem] in uno mundo magnitudines stellarum, situm et motum praeponderans et stellarum distantias taliter ordinans, ut, nisi quaelibet regio ita esset sicut est, nec ipsa esse nec in tali situ et ordine esse nec ipsum universum esse posset; dans omnibus stellis differentem claritatem, influentiam, figuram et colorem atque calorem, qui claritatem concomitatur influentialiter, et ita proportionaliter partium ad invicem proportionem constituens, ut in qualibet sit motus partium ad totum…'

3.5 Unification of the Realms and the Possibility of Life

One of the most revolutionary features of Gilbert's philosophy, which pervades much of *De mundo* as well as *De magnete*, was his attack on the traditional Aristotelian division between the sublunary and celestial realms. This attack is central to everything discussed thus far in terms of celestial influence and motion. Early on in *De mundo* he criticised philosophers who foolishly try to divide the world into the celestial and the elemental, and who claim the heavens to be made of incorruptible ether while the earth is sordid and mutable (Gilbert 1651, 8). Gilbert explained that 'it is ridiculous to divide the sensible world in such a way that within it there can be no comparison or likeness'.[61] It seemed clear to him that both space and nature are the same above the course of the moon and within it; there is no arbitrary divide. 'Bodies form above the moon and pass away', he claimed, 'they have birth and death, constancy and motion.'[62]

A significant portion of *De mundo nostro sublunari* was dedicated to a discussion of *luna* itself, which Gilbert called the partner of the earth (*telluris socia*) (Gilbert 1651, 170). He began, as he did in many of his chapters, by reviewing previous opinion on the subject matter. Some, like Orpheus, Thales, Pythagoras, Anaxagoras and Democritus, considered it to be another earth, possibly inhabited with animals, or even with its own cities. Others considered it to be made of fire, or air. The Peripatetics, of course, considered it to be made of the fifth essence.[63] Gilbert claimed that these opinions should not be accepted or rejected until after the observable features of the moon have been clearly expounded. Only then can one determine 'what the moon is'.[64] Soon afterwards he made that determination clear. 'We consider the moon to be another small earth', he wrote, 'or a body ordered in the way of the earth.'[65] According to Gilbert the similarities are numerous:

> The moon, different in nature from the sun, is a solid substance without light, diverse in its outermost parts, not condensed from an imaginary fifth essence, such that it is a thicker part of a nugatory sphere; rather it is a star, like the earth, having motion in its own space.[66]

This motion, being the moon's monthly orbit and rotation, serves the same purpose as the motion of the earth, that is, in order to enjoy the variety of beneficial

---

[61] Gilbert 1651, 9: 'quod ut vanum est et falsum, ita mundum sensibilem sic distinxisse, ut nulla esset in eo comparatio aut proportio, ridiculum'.

[62] Gilbert 1651, 155: 'Concrescunt enim supra Lunam corpora, et evanescunt, ortum habent et interitum, constantiam et motum.'

[63] Again, Gilbert paraphrased this summary of opinions from Patrizi. Compare Gilbert 1651, 170–71 with Patrizi 1591, fols 112^{r-v}.

[64] Gilbert 1651, 171: 'Nos vero, rejectis aut approbatis hisce opinionibus, tum demum, quid sit Luna, determinabimus, posteaquam plurima in ea. observanda, ostensa et demonstrata fuerint.'

[65] Gilbert 1651, 173: 'Quod vero Lunam, tellurem alteram minorem, aut corpus aliud telluris modo ordinatum, existimamus, id postea confirmabimus.'

[66] Gilbert 1651, 173: 'Luna, diversa natura a Sole, est solida absque lumine substantia, diversa in eminentiis, non provenit unquam ab inspissatione imaginatae quintae essentiae, adeo ut densior pars sit sphaerae nugatoriae; sed astrum est, sicut tellus, suis spatiis motum habens.'

influences from the sun and the stars (Gilbert 1651, 156–57). Gilbert had made a similar assertion in *De magnete* when comparing the reasons for the earth's motion with that of the other 'Wanderers'. The moon turns 'to receive in succession the sun's beams in which she, like the Earth, rejoices, and is refreshed'.[67]

The moon and the earth were truly companions in Gilbert's eyes. The earth acts for the moon as the moon does for the earth, supplying reflected light as compensation during the long lunar nights. The moon has continents and seas, the former showing as dark spots and the latter shining brightly (Gilbert 1651, 176–78).[68] Gilbert even included in *De mundo* one of the first ever selenographs, a map of the moon's surface drawn from the naked eye. He labelled the land masses and oceans, even granting the name *Britannia* to one of the smaller islands.[69] He disagreed with the belief that the dark and light patches on the moon are only areas of lesser and greater transparency, singling out Benedetti as one who held this opinion (Benedetti 1585, 299; Gilbert 1651, 173). This theory had agreed with Benedetti's belief that the moon and the planets acted as mirrors for their respective opaque earths (see Sect. 2.5), but didn't fit with Gilbert's attempt to establish the similarities between the primary globes. Nor did it suit his ideas about celestial influence.

Gilbert spent much of *De magnete* attempting to prove that the entire earth had a magnetic potency, and that this was indicative of an animate nature. This characteristic—of a soul capable of effusing its form—he then granted to the moon and all the other globes in the universe. He was keen to distinguish this from the 'formal cause of the Peripatetics', or the 'primary form of Aristotle'. Gilbert saw it as a 'primary, radical and astral form', inherent to the primary spheres of the universe and all their homogeneous parts.[70] In *De magnete* and *De mundo* the entire universe is animated, with the earth, the planets and every star possessing their own soul. These souls or astral forms are not all the same. Some, such as those of the sun and certain stars, are far more eminent and vigorous than others. Yet they all share some characteristics, such as the power to draw their parts back to themselves and extend their form outwards (Gilbert 1600, 208). Some, such as the earth and the moon, share more in common, as he explains in *De mundo*:

> Every one of the stars is imbued with its own primary and specific nature. For if in a field or meadow there are different types of plants, endowed with various virtues, so too are the great orbs informed by their own peculiar virtues; they are not idle lights, nor do they pour all their vigour into the earth. It is reasonable to think with the ancient philosophers that the

---

[67] Gilbert 1600, 224–25: 'Nam et Luna etiam menstruo cursu convertit sese, ut solis lumina successive recipiat, quibus non aliter atque tellus gaudet, et recreatur…'

[68] On Gilbert and the moon, see also Roos 2001, 88–93.

[69] The map is between two pages, Gilbert 1651, 172–73. See Pumfrey 2011.

[70] Gilbert 1600, 65: 'Forma illa singularis est, et peculiaris, non Peripateticorum causa formalis, et specifica in mixtis, et secunda forma, non generantium corporum propagatrix; sed primorum et praecipuorum globorum forma; et partium eorum homogenearum, non corruptarum, propria entitas et existentia, quam nos primariam, et radicalem, et astream appelare possumus formam; non formam primam Aristotelis, sed singularem illam, quae globum suum proprium tuetur et disponit.'

> moon, which consents with the Earth in many ways, and endures similar affections from the sun, is indeed another Earth.[71]

Thus, it is misleading to argue, as Freudenthal does, that Gilbert 'postulated only two cosmologically significant forces, the magnetic and the quasi-electrical,' and that 'according to his theory, the earth and all other celestial bodies (including the sun) are physically entirely equivalent and indistinguishable' (Freudenthal 1983, 32). Gilbert in fact placed a lot of emphasis on variety:

> The simple and uniform Earth is not so externally, just as the moon, or the planets, or the stars themselves are composed of different substances and are vastly different bodies. Some are clear, bright and shining, like the sun, Venus, Jupiter, Canis Major, Lyra; others are less splendid, like Saturn, Mercury, Gorgon, Corvus; other are wholly without light, such as the Earth, the moon, Praesepe, Sagittarius, Capricorn, the head of Orion.[72]

In Gilbert's opinion, the stars and planets are a mixed bag of different globes, some shining with their own light, others reflecting light, all with their own effluvia, some more dense than others. Gilbert filled his universe with a diverse range of bodies, and a corresponding diversity of 'forces'. Magnetism is the earth's astral form, sharing similarities with, but not identical to, those of the other globes.

Gilbert's aim was to completely remove the philosophical and *ontological* barriers that differentiated the earth from the other planets. 'Let this therefore be seen as a monstrosity in the Aristotelian universe', he declared in *De magnete*, 'in which everything is perfect, vigorous, animate; while the earth alone, an unhappy portion, is paltry, imperfect, dead, inanimate, and perishable'.[73] However, the other side of the coin for Gilbert was his conviction that there was variety and change in the heavens just as there was on earth. The stars are seen as varying in size and brightness, the light of the most excellent ones distinguished by the degree of scintillation. Some have dense effluvia which reflect the light from the sun. The Milky Way he described as real material, spread and settled through the void of the world, but discrete from that void. Of the planets, he seemed unsure whether some shine with their own light or whether they all must borrow it from the sun.[74]

This variety did not just exist between the different globes, but within the globes themselves. This was one of the most interesting repercussions of Gilbert's attack

---

[71] Gilbert 1651, 174: 'Unumquodque astrum sua specifica natura et primaria imbuitur. Nam si in prato et herbaceo campo, variae sint herbae, virtutibus variis dotatae, consentaneum est magnos orbes, suis etiam peculiaribus virtutibus informatos; non lumina sunt otiosa, non omnem vigorem in tellurem infundunt. Lunam, quae in plurimis cum tellure consentit, quae similibus affectionibus a Sole patitur, tellurem etiam alteram esse, cum antiquis philosophis putare consentaneum est…'

[72] Gilbert 1651, 110–11: 'Simplex terra et uniformis, licet non exacte, in externis; ut Luna, ut Planetae, ut Stellae etiam ipsae, ex varia substantia constant et longe diversa sunt corpora: unde alia clara, limpida, lucentia, ut Sol, Venus, Iupiter, Canis major, Lyra; alia minus splendida, ut Saturnus, Mercurius, Gorgon, Corvus; Alia prorsus absque luce, ut Terra, Luna, Praesepe, Sagittarius, Capricornus, caput Orionis.'

[73] Gilbert 1600, 209: 'Monstrum igitur istud in Aristotelico mundo videatur, in quo omnia perfecta, vivida, animata; unica vero terra, infoelix pars pusilla, imperfecta, mortua, inanimata et caduca.'

[74] Gilbert 1651, 206–10. On the Milky Way, see ibid., 52: 'Lacteus circulus est realis materia, fusa constansque per inania mundi, discreta tamen ab inani.' See also ibid., 247–50.

on Aristotelian cosmology: generation and corruption were not limited to the earth itself, but occurred also on the surfaces of the other planets and stars:

> All of these, as the rest of the stars, are said to be simple in their interiors and intimate parts. Those who are in admiration of such a great simplicity and, as they call it, fifth essence, are greatly fooled. The earth is in its own nature simple, true, homogeneous and uniform, yet its higher and exposed parts are shaken by solar lights as well as the lights of the other stars, and changed into the many species of things and corruptions. This is the origin of the varieties of animals, trees, plants, fossils; here is the origin and flowing of springs, seas and rivers.[75]

The illusion that the heavens do not change, or that their surfaces are composed of simple elements like ether or fire, is a product simply of their distance from an observer on earth. 'If an eye were carried above the moon', he asked in *De mundo*, 'would it observe changes in the plants of the earth, or the generation and corruption of other things?'[76] He expanded on this point again in the first book of the *Meteorologia*, where he took issue with an argument from Aristotle:

> 'In all previous times, as far back as records go, it can be understood that no change has been seen to ensue in the highest part of the heavens or any part of it.' It is now possible to prove, by various examples and reasons, the falsity of this. For the heavens are changed, and are ornamented with an admirable variety of things and vicissitude in their globes, in the same way as our earth appears to be. And yet no sense, no keenness of sight can see it. Just as if a human eye placed in the moon were to look down on the earth, it would not see what men have built, nor the delights of forests or plains, nor mountains, nor clouds or rivers. Rather it would see a globe which resembles the moon, reflecting the rays of the sun, marked with a mottled face and certain blemishes.[77]

We have already seen both Cusa and Benedetti advocate this kind of thought experiment, encouraging the reader to imagine what the earth might look like from a distance. Patrizi argued in an even more similar vein that the earth would appear much

[75] Gilbert 1651, 111: 'Quae omnia, ut reliqua etiam astra suis interioribus et intimis partibus simplicia dici possunt. Qui vero tantam supernorum corporum simplicitatem, et quintam, quam vocant, essentiam admirantur, multum illi decipiuntur. Terra est in sua natura simplex, vera, homogenica et uniformis, etiamsi priores et extimae partes a solaribus, nec non aliorum astrorum luminibus, labefactae sunt, et in multas species rerum et corruptelas immutatae. Hinc animalium, arborum, herbarum, fossilium varietates: fontium; marium, et fluviorum primordia et confluentiae.'

[76] Gilbert 1651, 176: 'Si enim supra Lunam oculus foret, an discerneret mutationem telluris in vegetabilibus, aliisque generationibus, et corruptionibus?'

[77] Gilbert 1651, 241–42: 'Omni enim praeterito tempore, quantum ex memoria, quam alii ab aliis traditam accepimus, intelligi potest, nulla aut in toto coelo summo, aut in ulla eius parte, consecuta videtur mutatio quod nunc minus verum esse variis exemplis et rationibus probari potuisset: tamen immutari coelum, et admirabili rerum varietate, et vicissitudine in suis globis (quemadmodum apud nos terra nostra) ornari, et tamen sensus noster nullus, nulla oculorum acies illam percipere potest. veluti si in Luna positus humanus oculus terras despectaret, non agnosceret quid facerent homines, non sylvarum, non camporum amoenitates, non montes, non nubes, non pluvias; globum tamen instar Lunae, Solis radios remittentem, varia facie et quibusdam quasi maculis insignitum, videret.' See Aristotle, *On the Heavens*, I.3, 270b12–17.

like the moon.[78] Gilbert had already demonstrated that that particular globe possessed land masses, oceans, mountains, etc. These passages suggest that these are features shared by the rest of the globes, and that their similarity to the earth extends to the possibility of life on their surfaces. A better understanding of this facet of Gilbert's cosmology can be achieved by examining his theories on the phenomenon of life itself, and generation more broadly.

'We consider life to be nothing other than the activity of the attenuated humour within the structures (*cancelli*) of the form.'[79] He stated this in *De mundo* while discussing the nature of heat. Later on, while arguing that the earth (and the rest of the globes along with it) possesses a soul, he defined the soul, in Aristotelian terms, as 'the activity of an organic body'.[80] Gilbert, who we must remember was a physician, opposed established medical theories of natural attraction, especially the notion that heat is an attractive force (Gilbert 1651, 98). This opposition went hand in hand with, or was perhaps necessitated by, his reduction of the elemental scheme and his dismissal of the active role of 'qualities'. He explained life and conception thus:

> We consider the invigorating force or soul to be a formal and organic activity, and the heat of plants and animals to be an activity of the attenuation of humours. Life, though, is a combined activity of heat and the soul; whether you prefer to call it soul, or the operative substantial form, life results from either force. The substantial form cannot remain or have effect without heat, such that the form in seeds will lie dormant and not grow until heat is moved through the sap, at which point the appropriate humour is expanded by attenuation, and flows through itself, emitting roots or appropriate instruments of nutrition, and grows larger and rises.[81]

But how did this concept of life agree with Gilbert's theory that everything on earth derived from the one true earth element? For Gilbert, true earth not only contains an intrinsic moisture or sap, it also contains the seeds of all variety on earth. For example, in *De magnete* he talked about how the earth produces the different metals. He disagreed with the notions of the alchemists (*chemistae*) that it has anything to do with quantity or proportion, or any material cause whatsoever. Rather, when beds or

---

[78] Patrizi 1591, fol. 112v: 'Neque ullum dubium nobis est, si quis nostrum, in lunam ascenderet, terramque inde prospectaret, eadem illi in terra apparitura, quae nobis hinc in luna apparent.' Translation: 'Nor do we doubt that, if one of us were to ascend to the moon and from there view the earth, the same things would appear on earth to him as appear to us on the moon from here.'

[79] Gilbert 1651, 81: 'nosque vitam nihil aliud esse volumus, quam actum humoris attenuati intra formae cancellos'.

[80] Gilbert 1651, 176: 'Anima est actus corporis organici.' See Aristotle, *On the Soul*, II.1, 412a30.

[81] Gilbert 1651, 97: 'positum enim est a nobis, vegetatricem vim sive animam, actum esse formalem organicum; caloremque stirpium, ut et animalium esse actum attenuationis humoris. vita vero est combinatus actus caloris et animae; sive animam volueris, sive substantialem formam operantem, per utrum vigorem vita existit. Nam et substantialis forma absque calore idoneo sopita diu manet, nec effectum producit, quemadmodum in seminibus cernere licet, quae non germinant, nec incrementum habent, donec calenti succo perfusa moventur. cum vero idoneus humor attenuando solvitur, fundunt sese, radicesque emittunt, tanquam nutritioni instrumenta idonea, germinant amplius et assurgunt'.

regions agree appropriately with the material or moisture or vapour from the earth, the metals 'assume forms from the universal nature, by which they are perfected; no different from the other minerals, plants, and animals whatever'.[82]

The seed of everything, animal, vegetable, mineral, is present in the true earth itself, awaiting the proper circumstances to manifest. This he explained further in *De magnete*:

> For the hidden beginnings of metals and stones lie inside the earth, as those of herbs and plants do in its outer crust (*peripheria*). For the soil dug out of a deep well, where would seem to be no suspicion of harboured seed, when placed on a very high tower, produces green plants and unbidden grass by the incubation of the sun and sky (*coelum*); and those of the kind which grow spontaneously in that region, for each region produces its own herbs and plants, also its own metals.[83]

This example of somewhat spontaneous generation was repeated in *De mundo*, along with further explanations of the role of the earth in generation:

> The earth is our common mother, it alone provides matter, and in it lie the seedbeds of things, which, when they are conceived in the fitting place by the activity of the humour, begin to have motion, change and condition from the activity of the humour within the structures of the form… Every single thing exists in the appropriate form of being. Those things which take root adhere to the earth and draw from it the sap necessary to its form… However, those things which live in the water move in the earth's thinner sap, and are preserved by the appropriate warmth of the water. The rest of the animals live in the air, which is the most attenuated fluid. They are rejuvenated by this spirit-like fluid, they prolong their life by its heat, and they consume food so that, the humour having been consumed and put in motion, they grow.[84]

In Gilbert's philosophy, the seeds lie dormant in the earth until germinated by moisture and heat. They grow by the addition of matter or humours—the appropriate nutrition being selected by the form—and this growth is directed again by the form into the appropriate shape (Gilbert 1651, 217–18). This process works similarly in animals as well as plants. In 1953, Jane Oppenheimer drew attention to an overlooked experiment in *De magnete* to do with plant grafting. In her opinion, the

---

[82] Gilbert 1600, 20: 'sed quando cum idonea materia alveis convenientibus, et regionibus, formas apprehendunt, ab universali natura, quibus perficiuntur; non aliter atque reliqua fossilia, vegetabilia, et animalia quaeque'.

[83] Gilbert 1600, 21: 'Latent enim in tellure metallorum et lapidum abdita primordia, ut in peripheria, herbarum et stirpium. Terra enim ex profundo puteo eruta, ubi nulla suspicio concepti seminis esse videatur, si in altissima turri posita fuerit, herbam producit virentem, et iniussa gramina, Sole et coelo terrae incubantibus; atque illa quidem quae in illa regione sunt spontanea; suas enim unaquaeque; regio herbas producit, et stirpes, sua etiam metalla.'

[84] Gilbert 1651, 46: 'Tellus communis mater est, haec sola materiam suppeditat, in ea. latent seminaria rerum; quae ut concepta fuerint loco idoneo, ab actu humoris, intra formae cancellos principium habent motus, augmentum, et statum. Elementa non aliunde petuntur: unaquaeque res in convenienti essentia; quae radices agunt, in tellure haerent, succumque inde hauriunt cuique formae necessarium. Succus ex tellure ducitur, quae undique et per universam molem succum habet insitum et genuinum. Quae vero in aquis degunt, in telluris solutiori succo vagantur, et aquarum convenienti tepore conservantur. In aere vero, qui fluor est attenuatus magis, caetera degunt animalia, fluore spirituali recreantur, eiusque etiam tepore vitam producunt, alimentaque etiam assumunt, ut acto et absumpto humore nova addantur incrementa.'

purpose of the experiment, seemingly out of place in a work on magnetism, was to demonstrate the importance of 'form' in the growth of both mineral and living substance (Oppenheimer 1953). It is interesting to note that throughout all these processes—through the growth, development and change of living things—the hypostasis of the true earth remains embedded in the nature of everything.[85]

Gilbert considered the life of all animals and plants to depend upon the living nature of the globes themselves, and thus celestial influence is intimately connected with his concept of an animate universe. In *De magnete*, when discussing the souls of the globes, he admitted that they would not have organs like other animals, 'yet they live and imbue with life the small particles in the outer parts of the earth'. Furthermore, 'since living bodies arise and receive life from the earth and the sun, and plants grow on the earth without any seeds thrown down… it is not likely that they can produce what is not in them; but they awaken life, and therefore they are living'.[86] We need not look to Bruno as a source for this kind of animism. After all, we saw the very same argument put forward by Averroes (Sect. 2.2). Like the motion of the globes, which is a composite of interior and exterior forces, life too is the combined product of the innate seeds and moisture of the earth with the life engendering virtues of the sun and other globes.

Gilbert never discussed the nature of life on other planets or stars, nor did he explicitly refer to there being animals or rational creatures on them. He wasn't, like Bruno, the champion of an infinitely populated universe. Nor did he speculate on the nature of extraterrestrial beings, as Nicholas of Cusa had done. Let us examine what he did commit to. In terms of being composed of a homogeneous substance and having a unique and extended astral form, the earth and the other stars and planets are of a kind. They also share the same reasoning behind their motions, i.e., the regulation of celestial influence to aid the conservation, perfection and ornamentation of their parts. The significance of this for the production of living things, at least in terms of the earth, is made clear in *De magnete*:

> But it is ridiculous for a philosophical person to suppose that all the fixed stars and the planets and the still higher heavens revolve to no other purpose, except the benefit of the Earth. It is the Earth, then, that revolves, not the whole heaven, and this motion gives opportunity for the growth and decrease of things, and for the generation of living things, and awakens an internal heat for the bearing of life.[87]

---

[85] Gilbert 1651, 109: 'Sed in terra firma et magnetica humor insitus exspirare potest; tota vero natura in aquam non solvitur, manet enim in rerum natura hypostasis telluris immutabilis.' Translation: 'The intrinsic humour in the firm and magnetic earth can be exhaled, but the whole nature is not changed into water. Instead, the immutable hypostasis of earth remains in the nature of things.'

[86] Gilbert 1600, 209: 'vivunt tamen, et vita imbuunt corpuscula in terrenis eminentiis… Cum vero a tellure et sole viventia corpora oriantur et animentur, crescantque in terra herbae absque ullis iactis seminibus … non verisimile est posse illa efficere quod in illis non sit, sed animas excitant, ideoque sunt animata'.

[87] Gilbert 1600, 225: 'Volvi vero astra omnia fixa, et errones, caelosque adhuc superiores, necquiquam nisi telluris commodo, homini philosopho ridiculum est putare. Volvitur igitur tellus non caelum totum; qui motus incrementis et decrementis rerum, et animantium generationi occasionem adfert, et intestinos calores ad foeturam excitat.'

At the end of Book 5 of *De magnete* is one of the clearest indications in his work that he considered the other planets to be inhabited. All the globes are living and each possesses a soul of its own, as we have seen, which pervades its interior and homogeneous parts, but ‘it is only on the surfaces of the globes that the multitude of living and animated beings is clearly perceived, by which great and delightful variety the great maker is well pleased.’[88] He wasn't referring here just to the earth but rather to the ‘globes’ plural. If we couple this with his arguments in *De mundo* that the heavenly bodies have their own effluvia, experience change, and enjoy admirable varieties and vicissitudes like the earth, it becomes clear that life on the surface of other globes was at the very least a possible, and more likely a realised, feature of Gilbert's cosmology. According to Gilbert the earth is a planet, or rather the earth, planets and stars are animate globes, distinct but of a kind. This fundamental tenet combined with Gilbert's understanding of planetary motions as self-regulatory in terms of light and celestial influence. These influences are not only life *sustaining* but life *engendering*, giving rise to the conclusion that the surfaces of the moon and the other planets are adorned with a variety of living things, as the earth is.

3.6 Conclusion

Unlike others that followed him, or indeed some that preceded him, Gilbert did not expand upon the subject of living beings on other planets. His chapters on the moon in *De mundo*, including his selenograph, are the only example of Gilbert investigating the nature of the surface of any other globe. The importance of his philosophy lies in the removal of any natural barriers to the existence of ET life. Crucially, in regard to our present investigation, this did not preclude a belief in astrology. In fact, astrology was central to his extension of life to the other globes. He represents further evidence that the development of pluralist ideas was often linked to changing ideas about the function, nature and extent of celestial influence. For Gilbert, celestial motion could be explained as individual globes endeavouring to regulate influences in order to preserve habitable conditions on their exteriors. The predominant influence is the light from the sun, but it also includes the more particular effects of the planets and stars. The celestial bodies move not for our benefit, but for their own, a theory which only makes sense if their surfaces are habitats like that of the earth. Gilbert thus united the terrestrial and celestial realms, theorising an astrologically vibrant universe throughout which life thrives in its many forms.

It is unclear how many people read *De mundo*, either in its manuscript form or after it was eventually published nearly 50 years after Gilbert's death.[89] His legacy was guaranteed by *De magnete*, and his attribution of magnetism as the method of

[88] Gilbert 1600, 210: ‘Sed in globorum extremitatibus tantum, animarum et animatorum frequentia manifestius cernitur, in quibus summus opifex, maiore et iucunda varietate sibi perplacet.’

[89] Two known readers were Francis Bacon (1561–1626) and Thomas Harriot (c. 1560–1621). To this can be added Otto von Guericke, as will be discussed in Chap. 5.

the earth's rotation, as well as his comparison of the earth to the other planets, opened up the door for the application of terrestrial or sublunar philosophy to the celestial realm.[90] Gilbert's new philosophy is deserving of a more prominent place in the history of the ET life debate, both because of the obvious influence of pluralist ideas on his cosmology, and because of the influence that his theories had on subsequent philosophers and astronomers. His description of magnetism as the astral virtue of the earth should also be of interest to historians of astrology, as it represents a bridge between older conceptions of astrological virtues and newer ideas about celestial physics. The link between astrology and pluralism makes him particularly relevant to the history of astrobiology. His ideas about matter theory, magnetism, biology and planetary motion all combined in a unique vision of life in a cosmic context, or perhaps of the cosmos in a vital context. In the next chapter, we turn to one person who was greatly influenced by Gilbert, and who earned notoriety for his ideas about both celestial physics and ET life: Johannes Kepler.

References

Barlow, William. 1616. *Magneticall aduertisements, or, diuers pertinent obseruations, and approued experiments concerning the nature and properties of the load-stone*. London: Timothy Barlow.

Benedetti, Giovanni Battista. 1585. *Diversarum speculationum mathematicarum et physicarum liber*. Turin: Heirs of Nicola Bevilaqua.

Bennett, J.A. 1981. Cosmology and the magnetical philosophy 1640–1680. *Journal for the History of Astronomy* 12: 165–177.

Calcagnini, Celio. 1544. *Opera aliquot*. Basel: Hieronymus Frobenius.

Curry, Patrick. 1989. *Prophecy and power: Astrology in early modern England*. Princeton: Princeton University Press.

Cusa, Nicholas of. 1932. *Opera omnia*. Vol. 1. Leipzig: Felix Meiner.

Danielson, Dennis Richard. 2006. *The first Copernican: Georg Joachim Rheticus and the rise of the Copernican Revolution*. London: Walker.

Dunn, Richard. 1994. The true place of astrology among the mathematical arts of late Tudor England. *Annals of Science* 51: 151–163.

Freudenthal, Gad. 1983. Theory of matter and cosmology in William Gilbert's *De magnete*. *Isis* 74: 22–37.

Gatti, Hilary. 1999. *Giordano Bruno and renaissance science*. Ithaca/London: Cornell University Press.

Gaukroger, Stephen. 2002. *Descartes' system of natural philosophy*. Cambridge: Cambridge University Press.

Gilbert, William. 1600. *De magnete, magneticisque corporibus, et de magno magnete tellure: Physiologia noua, plurimis et argumentis, et experimentis demonstrata*. London: Peter Short.

———. 1651. *De mundo nostro sublunari philosophia nova*. Amsterdam: Lodewijk Elzevir.

[90] On the magnetic philosophy in the seventeenth century, see Bennett 1981; Pumfrey 1987; Pumfrey 1990. See also Wang 2016, although Wang's assertion that Gilbert didn't believe, or encourage others to believe, in action at a distance is debatable. On Gilbert's influence on the astronomer David Origanus (1558–1629), see Omodeo 2014, 152–53.

Henry, John. 2001. Animism and empiricism: Copernican physics and the origins of William Gilbert's experimental method. *Journal of the History of Ideas* 62: 99–119.

Hooykaas, R. 1984. *G. J. Rheticus' treatise on holy scripture and the motion of the Earth: With translation, annotations, commentary, and additional chapters on Ramus-Rheticus and the development of the problem before 1650*. Amsterdam/New York: North Holland.

Hutchison, Keith. 1982. What happened to occult qualities in the scientific revolution? *Isis* 73: 233–253.

Kargon, Robert Hugh. 1966. *Atomism in England from Hariot to Newton*. Oxford: Clarendon Press.

Koyré, Alexandre. 1957. *From the closed world to the infinite universe*. Baltimore: Johns Hopkins Press.

Lindberg, David C. 1986. The genesis of Kepler's theory of light: Light metaphysics from Plotinus to Kepler. *Osiris* 2: 4–42.

Omodeo, Pietro Daniel. 2014. *Copernicus in the cultural debates of the Renaissance: Reception, legacy, transformation*. Leiden: Brill.

Oppenheimer, Jane M. 1953. William Gilbert: Plant grafting and the grand analogy. *Journal of the History of Medicine and Allied Sciences* 8: 165–176.

Patrizi, Francesco. 1591. *Nova de universis philosophia*. Ferrara: Benedetto Mammarello.

Pumfrey, Stephen. 1987. *William Gilbert's magnetic philosophy 1580–1684: The creation and dissolution of a discipline*. PhD thesis, University of London.

———. 1990. Neo-Aristotelianism and the magnetic philosophy. In *New perspectives on Renaissance thought: Essays in the history of science, education and philosophy: In memory of Charles B. Schmitt*, ed. John Henry and Sarah Hutton, 177–189. London: Duckworth.

———. 2002. *Latitude and the magnetic earth*. Cambridge: Icon.

———. 2004. Gilbert, William. In *Oxford dictionary of national biography*, ed. H.C.G. Matthew and Brian Harrison, vol. XXII, 195–202. Oxford: Oxford University Press.

———. 2011. The *selenographia* of William Gilbert: His pre-telescopic map of the moon and his discovery of lunar libration. *Journal for the History of Astronomy* 42: 193–203.

Ridley, Mark. 1617. *Magneticall animadversions*. London: Printed by Nicholas Okes.

Roller, Duane. 1959. *The De magnete of William Gilbert*. Amsterdam: Hertzberger.

Roos, Anna Marie Eleanor. 2001. *Luminaries in the natural world: The sun and the moon in England, 1400–1720*. New York/Oxford: Peter Lang.

Rosen, Edward. 1939. *Three Copernican treatises*. New York: Columbia University Press.

Wang, Xiaona. 2016. Francis Bacon and magnetical cosmology. *Isis* 107: 707–721.

Westman, Robert S. 1975. The Melanchthon circle, Rheticus, and the Wittenberg interpretation of the Copernican theory. *Isis* 66: 165–193.

Whewell, William. 1857. *History of the inductive sciences from the earliest to the present time*. Vol. 3. 3rd ed. London: J. W. Parker.

Chapter 4
Johannes Kepler: A New Astronomy, Astrological Harmonies and Living Creatures

> *Thus the harmonies of music are sought within by the singer; the harmonies of the rays are looked for outside by sublunary Nature, are observed when met, are discriminated from those which are not harmonic (and thus take from it their essence), are selected, and are applied. In brief, the configurations sing the leading part; sublunary Nature dances to the laws of this song.*
>
> (Kepler 1997, 325)

> *Thus this appearance, brought by the agency of light to the body of the Sun, can along with the light itself flow straight to living creatures, who share in this instinct, just as in the fourth book we have stated that the pattern of the heaven flows to a foetus by the agency of the rays.*
>
> (Kepler 1997, 424)

Abstract Johannes Kepler is a key figure in both the histories of astrology and the extraterrestrial life debate. To date, however, historians have not appreciated the intimate connection between the two concepts in Kepler's thought. This chapter presents a thorough reconstruction of Kepler's astrological theories and an analysis of his various writings on pluralist themes, attempting to establish exactly what fuelled his commitment to the existence of extraterrestrial life and how it fitted into his highly theological cosmos. As Gilbert's had done, Kepler's astrological theories expanded the operation of celestial influence to centres other than the earth, and his teleological understanding of celestial influence necessitated the existence of observers (i.e. living creatures) at those centres.

Keywords Johannes Kepler · Astrology · Extraterrestrial life · New star · Harmony · *Somnium*

© Springer Nature Switzerland AG 2019

J. E. Christie, *From Influence to Inhabitation*, International Archives of the History of Ideas Archives internationales d'histoire des idées 228,
https://doi.org/10.1007/978-3-030-22169-0_4

4.1 Introduction

Johannes Kepler (1571–1630), the imperial mathematician who discovered the elliptical orbits of the planets, and whose name has recently been given to the thousands of exoplanets discovered by the Kepler orbital telescope, was also one of the last great minds to lend his talents to the scientific justification of astrology. His astrological theories, almost entirely based upon the influence of harmony on the soul, were as unprecedented as they were un-replicated. Often seen as an example of his Janus-like nature, facing both the ancient and modern worlds, Kepler's astrology was in fact a crucial part of the cosmology which gave birth to his now vaunted laws of planetary motion. At the same time, his contributions to the pluralist philosophy 'make one of the most dramatic stories in the entire history of the extraterrestrial life debate' (Crowe 1986, 10). Throughout his career Kepler vocally advocated the existence of living creatures on the other celestial globes, including the moon, the planets, and eventually even the sun itself. It seems that initially this advocacy was based upon perceived similarities between the earth and the other globes, in particular the earth-like geographical features of the moon. Yet this is only one part of the story. In fact, Kepler's astrological beliefs provided an even more compelling reason to assume the existence of ET life. His astrological theories expanded the operation of celestial influence to centres other than the earth. In particular, his discovery of astrological harmonies centred on the sun necessitated the existence of souls engaged in the active reception of that influence.

There is no space here to attempt to reconstruct a comprehensive account of Kepler's cosmology. Thankfully, the ever-increasing field of Kepler scholarship has produced some thorough studies of various aspects of Kepler's astronomy, astrology, physics and metaphysics.[1] This chapter will focus mainly on instances of Kepler's discussion of astrological and pluralistic themes, discussing other areas of his philosophy when relevant, to analyse exactly what fuelled his commitment to the existence of ET life and how it fitted into his highly theological cosmos. During the course of this chapter, comparisons will be drawn between Kepler and Gilbert, highlighting points of convergence and departure between the cosmologies of these two important figures in the early history of pluralism. In Kepler's cosmos, like Gilbert's, the celestial bodies serve as sources of influence while at the same time providing habitats for their own creatures. He therefore represents an intermediary stage in the transformation of astronomical teleology which saw astrology supplanted by pluralism. Kepler's extraterrestrials, like his laws of planetary motion, would eventually be separated from the philosophy in which they were conceived.

[1] Some of the more relevant works for this discussion are Simon 1975; Field 1984, 1988; Stephenson 1987, 1994; Rabin 1997; Methuen 1998; Martens 2000; Voelkel 2001; Kremer and Włodarczyk 2009; Boner 2013.

4.2 Kepler's Early Works: The Moon, Optics and Astrology

Kepler's first publication was his *Mysterium cosmographicum* ('The Cosmographic Mystery'), in which he constructed a defence of the Copernican system based on astronomical, physical, geometrical and teleological grounds.[2] He argued against the existence of solid spheres or orbs in the heavens, claiming rather that the region was full of what he called an ethereal or celestial 'air' (*coelestis aura*) (Kepler 1937–, I, 54). He also argued against the Aristotelian notion of gravity and levity, in which elements were supposed to seek their natural place, in favour of the Copernican notion (as it was now thought of) that everything of the same nature as a body sought its centre (Kepler 1937–, I, 59). For the first time as well, he proposed a motive soul or spirit in the sun which was responsible for driving the orbits of the planets (Kepler 1937–, I, 70). Yet quite aside from these seemingly modernising developments, the real crowning glory of this publication in Kepler's eyes was his theory that the number and spacing of the planets in the Copernican system were determined by the five Platonic solids (Kepler 1937–, I, esp. 26). With this theory, in which the regular solids (the tetrahedron, cube, octahedron, dodecahedron, and icosahedron) fit in between the insubstantial orbital spheres of the six planets, Kepler believed he had found the secret geometrical blueprint for God's creation of the universe.

This discovery, however, carried a significance beyond astronomy. These solids were not just quantitative, determining numbers and distances, but *qualitative*, relating to both the order and operation of the celestial world. In Kepler's system, the earth, situated between the dodecahedron and the icosahedron, created a separation between the two classes of solids. This fit with his conviction that the earth was the 'pinnacle and pattern of the whole universe' and the most important of the planets: the dwelling place of Man, God's favourite creature and the purpose of all creation (Kepler 1937–, I, 28). On a more practical note, Kepler also suggested in the *Mysterium* that the virtue or power of a planet, astrologically speaking, bore a relationship to the solid or solids which determine its orbit. 'I shall have the physicists against me in these chapters', he stated, 'because I have deduced the natural properties of the planets from immaterial things and mathematical figures' (Kepler 1937–, I, 37, 1981, 123). In fact, this practice of deducing qualitative properties from quantitative ratios would become a hallmark of Kepler's natural philosophy, and in this work we see his first attempt to give a rational basis for the efficacy of astrological aspects (Kepler 1937–, I, 43). Astrology was an example of how geometry provided the blueprint not only for creation but also for the continuing operation of nature, and it remained a source of practical and philosophical fascination throughout his life.[3]

---

[2] Kepler, *Mysterium cosmographicum* (1596), in Kepler 1937–, I, 1–80. English translations are from Kepler 1981.

[3] On this point, see especially Field 1988.

One other such source of fascination was provided by pluralism. Kepler's first treatment of pluralist themes appeared in the *Mysterium* during his discussion of the moon. In one particular passage, he suggested that the moon is dragged around by the earth (Kepler 1937–, I, 55–56), an early indication of the development of Kepler's celestial physics which would develop still further once he read Gilbert's *De magnete*. He then portrayed the moon as a source of influence; a steward serving its master, the earth, with light and moisture. He also asserted that an observer on the moon would be unsure of his direction or even if he is moving at all. This argument, designed to counter those against the motion of the earth, formed the core of the disputation that Kepler had written as a student, but been prevented from defending, at Tübingen in 1593. This work developed throughout his life before being posthumously published as his *Somnium* (1634).

In one line of particular relevance to the issue of pluralism, Kepler considered it as proven that the moon contained geographical features which correspond to those of the earth, and went on to say that 'on this account alone Copernicus is more convincing, as he endows these two bodies with a common position and motion' (Kepler 1937–, I, 56, 1981, 165). Kepler was encouraged to engage with Copernicanism by Michael Maestlin (1550–1631), professor of mathematics at Tübingen, who played an important role in Kepler's early lunar theories as well.[4] What is interesting is that both before and after this mention of the earth-like features of the lunar surface, Kepler linked it with the Copernican argument that they shared a common space and motion. By connecting the common nature of the earth and moon with their common situation in the Copernican scheme, he was in fact suggesting that the former was a consequence of the latter.

At this early stage in his career, it appears that Kepler only extended earth-like qualities as far as the lunar globe, and that as part of a larger attempt to build evidence for the Copernican cause. Part of this attempt is continued in his first work on optics, the *Astronomiae pars optica* ('The Optical Part of Astronomy') published in 1604. Kepler considered optics as the third part of astronomy, along with the mechanical part (instruments), historical part (observations), and the fourth 'physical part' (Kepler 1937–, II, 14). The first part of this work is laden with cosmogonic, metaphysical, and anatomical theories, some of which may be outlined here. Kepler believed that by giving the universe the shape of the sphere, God created in it an image of the Trinity, with the Father as the centre, the Son as the circumference, and the Holy Spirit as the intervening space (Kepler 1937–, II, 19). Material objects, such as the magnet, effused their virtues in a sphere. Light, as an effused virtue of the sun, proceeded in an infinite number of straight lines in an instant, or infinitely quickly, but was attenuated with breadth.[5] The sun, accordingly, must be in the centre in order to project its light equally into a sphere (Kepler 1937–, II, 19). Furthermore, Kepler assumed that all light had heat. Within the human body, the heart contained a flame which transmitted heat to the body through the arteries, and so animal heat depended on light. The soul itself had an essence similar to light, and

[4] See for example Grafton 1973; Caspar 1993, 46; Methuen 1996.

[5] On Kepler's metaphysics of light, and his debt to Plotinus, see Lindberg 1986.

their fellowship was analogous to that of heat with light, 'so far as light is the offspring of the soul.'[6]

Pluralist ideas again materialise in the chapter on spots on the moon.[7] 'I assert the body of the moon to be of such a kind as this our earth,' wrote Kepler, 'one globe made from water and continents.'[8] After presenting his reasons derived from optics to support this assertion, Kepler made his first mention of ET life:

> All these things provide evidence to what I have said: Plutarch is right in stating that the moon is like the earth, uneven and mountainous, and that, in proportion to its globe, its mountains are taller than the ones on earth. And likewise, allow me to play with Plutarch: since it comes to us by experience that men and animals take after the nature of their own land or region, therefore there will be living creatures (*creaturae viventes*) in the moon that are much greater in the size of their bodies and the hardness of their temperament than we are. The reason is clearly that, if there is anybody there, they endure days that are fifteen times longer than ours, with unimaginable heat, as the oppressive sun is overhead for a long time. Thus it was not rashly believed by common superstition to be the place appointed for the purification of souls.[9]

Here Kepler speculated that if there were living creatures on the moon their physiology would be adapted to the evident nature of their dwelling place. Just as their mountains are bigger in comparison to the moon's diameter, so their bodies would be larger than ours. They must also be tougher to endure the intense heat of the long lunar days. The concept of the celestial bodies as destinations for the soul after death is representative of that strand of Christian-Platonic eschatology, immortalised by Dante's *Paradiso*, which will continue to echo through Kepler's entanglement with pluralism, as indeed it has echoed through the 'plurality of worlds' philosophy up to the present day.[10] Even as Kepler appears to us to break new ground by introducing pluralist themes into the serious study of astronomy, he did so by presenting it through the lens of his classical and religious heritage.

---

[6] Kepler 1937–, II, 36: '…quatenus lux animae soboles'.

[7] Kepler again highlights the importance of lunar theory to the Copernican cause. See Kepler 1937–, II, 224: 'Tandem vero, ubi Plutarchus, ubi Moestlinus aequis in Philosophia auribus fuerint accepti: tum bene Aristarchus cum Copernico suo discipulo sperare incipiat.' Translation: 'Once Plutarch and Maestlin have been heard in philosophy with impartial ears, then Aristarchus and his disciple Copernicus may well begin to hope.'

[8] Kepler 1937–, II, 118: '…dicamque, Lunae tale esse corpus, quale haec nostra terra est, ex aqua et continentibus unum globum efficiens'.

[9] Kepler 1937–, II, 220: 'Haec omnia mihi praebent argumentum eius quod dixi: recte Lunam a Plutarcho tale corpus dici, quale terra est, inaequale montuosumque, et maiores quidem montes in proportione ad suum globum, quam sunt terreni in sua proportione. Ac ut cum Plutarcho etiam iocemur: quia penes nos usu venit, ut homines et animalia sequantur ingenium terrae seu provinciae suae: Erunt igitur in Luna creaturae viventes, multo maiori corporum mole, temperamentorumque duritie, quam nostra: sane quia et diem quindecim nostros dies longam, et ineffabiles aestus, Sole verticibus tam diu incumbente, perferunt, siquidem aliqui ibi sunt. Ut non absurde locus ille gentium superstitione lustrationi animarum destinari creditus sit.'

[10] In particular in religious movements such as Swedenborgianism and Mormonism. See Paul 1986; Dunér 2016.

There is no mention here of *why* precisely there should be creatures on the moon. The proposition is instead framed as a joke shared with Plutarch, whose *De facie* he had encountered since writing his student disputation.[11] But this joke, as is the case for most of Kepler's jocular outbursts, is in fact serious (Jardine 2009). The argument which is implicit here is made explicit in Kepler's second optical work, *Dioptrice* ('Dioptrics'), published in 1611 (Kepler 1937–, IV, 327–414), in which he revisited the optical part of astronomy in the wake of the recent application of the telescope:

> For if one wishes to focus the power of reason on these new observations, who does not see how far the contemplation of nature will extend its boundaries, while we inquire for whose good (*cui bono*) the tracts of mountains and valleys and the large expanses of oceans in the moon are, and whether a certain creature, not less ignoble than man, can be proposed, which inhabits those tracts.[12]

This *cui bono* argument lies at the heart of most debates on the topic of cosmological pluralism in the seventeenth century and beyond. We saw it already in Patrizi's repetition of the dictum that God does nothing in vain (Sect. 2.5). It is usually a consequence of the discovery of celestial objects or phenomena which conflict with the conception of the universe as centred around and geared towards the only inhabited place within it: the earth. Here we see the process of geocentrism and anthropocentrism giving way to a kind of anthropomorphism-*cum*-geomorphism, where a particular feature of the planetary or celestial realm is interpreted through analogy to its terrestrial equivalent and understood function.

Kepler's early engagement with pluralism, as mentioned, was limited to the moon. Obviously, of all the celestial globes it displays its imperfections, and subsequently its similarity to the earth, most clearly. But there is another possible reason. Before the publication of Galileo's telescopic observations, Kepler believed that the other planets were in fact different in nature to the earth and the moon in one important aspect: light. In his *De fundamentis astrologiae certioribus* ('On the More Certain Foundations of Astrology', 1601), a work on the principles of astrology attached to his prognostication for the following year, he took a somewhat agnostic stance on the issue of whether the planets only reflected the light of the sun or whether they had their own intrinsic light as well.[13] In his *Astronomiae pars optica* he took a more positive position and presented several arguments, derived from his

---

[11] Kepler's own Latin translation is in Kepler 1937–, XI.2, 380–483. See Appendix D in Kepler 1967, 209–211.

[12] Kepler 1937–, IV, 342: 'Quod si cui iam super novis hisce observationibus lubeat etiam Rationis vim excutere: quis non videt, quam longe contemplatio Naturae sua pomoeria prolatura sit; dum quaerimus, cui bono in Luna sint montium valliumque tractus, marium amplissima spacia; et an non ignobilior aliqua Creatura, quam homo, statui possit, quae tractus illos inhabitet.'

[13] Kepler, *De fundamentis astrologiae certioribus* (1601), in Kepler 1937–, IV, 7–35. On planetary light, see especially ibid., 17–20. See also the notes to these sections in the English translation: Field 1984, 236, n. 16, 243, ns 29, 30. The more traditional Aristotelian stance was that the planets are wholly illuminated by the sun. Grant traces the alternate hypothesis to Avicenna and Macrobius, with further arguments developed by Nicole Oresme and Albert of Saxony. See Grant 1994, 393–413, esp. 400–402.

astrological work, for why the planets shine with a 'double light' (Kepler 1937–, II, 228–29). Firstly, Venus does not seem to demonstrate phases like the moon. Secondly, not all light must come from the sun, as some things in the sublunary world have their own intrinsic light. Thirdly, he argued that from a certain combination of geometrical differences, and the *function* of celestial light, it is necessary that the planets have both their own light and that from the sun.[14] Lastly, he gave his reasons for the effects of the planets being a result of their nature, with colour being indicative of nature.[15]

In his *De fundamentis*, Kepler had argued for three main astrological 'causes'. The first was physical. The sun heats through direct light; the moon humidifies through borrowed light. The planets do both, as well as transmitting some of their own nature as indicated by their colour (Kepler 1937–, IV, 12–16). The second cause was geometrical, dealing mainly with aspects. The soul of the earth (more on this below), or the soul of a man, discerns when two or more planets are arranged at harmonic angles, and is at that moment roused into action. For the soul of the earth this means an increase in the production of vapours and therefore adverse weather (Kepler 1937–, IV, 23–24). This theory of the aspects, developed philosophically through the study of geometry and the earth's soul, and experimentally through the observation of the weather, formed the core of Kepler's astrological project. A third cause, expressed at this point only as a suggestion, is what Kepler called a 'harmonic cause' (Kepler 1937–, IV, 25–26). If the soul can recognise harmonic angles, why not also harmonic ratios between the orbital velocities of the planets? In his more mature astrology, Kepler would largely abandon the physical cause, at least with regards to the planets and the humidifying power of reflected light, and focus on developing the second and third causes. At this early stage, however, it seems that, along with a lack of observational evidence, the need to preserve astrological efficacy was part of the reason why Kepler maintained a diversity in nature between the earth and the moon on the one hand, and the planets on the other.

4.3 Kepler's Nova and Its Pluralist Significance

Not long after the completion of his *Astronomiae pars optica*, however, Kepler professed his belief in the wider diffusion of ET life through the universe. The occasion which prompted this response was the controversy over the new star of 1604. This supernova, now known as 'Kepler's Nova', aroused fascination and trepidation across Europe.[16] Large numbers of works of varying lengths and degrees of sophistication were published on the possible causes and astrological significance of the

[14] He didn't proceed to the conclusion that the earth, one of the planets, has its own light, as Cusa had done.

[15] The argument that the varying colours of the planets suggested a degree of self-luminosity was made by Albert of Saxony. See Grant 1994, 402.

[16] See Granada 2005; Boner 2007.

new star. Many called for Kepler, then the Imperial Mathematician, to publish his own views. He quickly published a short tract in German, but then delayed publication of a full treatment in Latin until 2 years later.[17] In *De stella nova* ('On the New Star'), published in 1606, Kepler was compelled to discuss a wide range of astronomical issues relevant to the appearance of the new star.[18] Long sections were dedicated to attacks on Aristotelian cosmology, the infinite universe of Bruno, and large swathes of astrological theory and practice.

When discussing what caused the nova and how it was created, Kepler tried at all times to refrain from a divine or astrological explanation. Instead, he maintained that the nova was created by a natural process with a terrestrial equivalent. The earth, he believed, has a natural faculty which produces new things out of excess material. These things include weather phenomena, gems and minerals, plants, and small animals. The problem of spontaneous generation, as we have seen, was an old one in philosophy, and much discussed in Renaissance medical humanism.[19] Kepler here aligned himself with some of the tenets of the Platonism of Marsilio Ficino, who believed the soul of the earth, a sub-class of the world-soul, to be responsible for the production of plants and small animals from waste materials.[20] Kepler argued that if the earth, one of the globes, possesses this faculty, then the other globes should possess it too. Yet the new star was produced not in a globe but in the ether itself, so it is necessary that there should be some similar faculty spread throughout the whole of the universe that cleans the ether by conglomerating and igniting excess material (Kepler 1937–, I, 269).

Kepler revealed his immediate source for this concept in a letter to David Fabricius (1564–1617), part of a detailed correspondence between the two men on the subject of the new star:

> I do not differ greatly from the philosophy of Cornelius Gemma, who supposes there to be one and the same spirit in the whole universe changing bodies (*somatomorphōn*) everyday, which acts according to the principle of the better and recognises what can be made most suitably anywhere from redundant material. Therefore, it changes the sweat of a woman or dog into head lice and fleas, dew into locusts and caterpillars, rope into eels, mud into frogs, water into fish, earth into plants, corpses into worms, dung into beetles and infinite new and unusual things; the exhalations of the air into shooting stars, and those of the ethereal region into comets and finally into stars. You can see that everywhere there is something moist that contains a seminal principle and produces the diversity of the species, so that not from the leaves of just any tree is a particular caterpillar born, but specific caterpillars from specific leaves. Yet that spirit of the whole world seems to exist for this purpose: that all things are regulated with respect to each other, and suitable bodily organs are added to a new creature. If nothing lived, it alone would enliven all matter, just as the earth, if nothing moved, would attract everything to itself, some things closer than others.[21]

---

[17] Kepler, *De stella nova in pede serpentarii* (1606), in Kepler 1937–, I, 149–356.

[18] See Boner 2011; Granada 2011a.

[19] For discussions on this topic, see Hirai 2011.

[20] See Hirai 2011, 124–25; Snyder 2011.

[21] Kepler to Fabricius (11 Oct 1605), no. 358 in Kepler 1937–, XV, 258: 'Ego vero non longe absum a Cornelii Gemmae Philosophia, qui existimat inesse unum et eundem spiritum in toto universo σωματομορφοῦντα quotidie, qui enim agat διὰ τὸ κάλλιον καὶ βέλτιον, et noverit, quid

Here we can compare the theories of generation of Kepler and Gilbert.[22] Both gave credit for spontaneous generation ultimately to the soul of the earth, and more immediately to a seminal principle in fluid matter. Celestial influence played a role for both philosophers. In Gilbert, the light acts upon and expands the humours, allowing the seeds to grow within the form seemingly placed upon them according to their situation on the earth (see Sect. 3.5). For Kepler, astral influence spurs on the soul of the earth in its somewhat more autonomous generative methods. At the end of the passage just quoted, Kepler compared the vivifying action of this universal spirit to the attraction or gravity of the earth. This demonstrates how interconnected all these ideas were both for Kepler and his contemporaries. The letter to Fabricius contained material from the drafts of both *De stella nova* and *Astronomia nova* (1609). Observe again the heavy emphasis Kepler put on purpose in regard to natural processes. 'This certainly is that faculty', he asserted in *De stella nova*, 'which everyday finds certain superfluous matter and converts it into the kind of small animal that is useful to nature, either in assisting or relieving.'[23]

Soon after the publication of *De stella nova*, Kepler's friend Johann Georg Brengger, who practised medicine in Kaufbeuren and Augsburg, took issue with his theory on the cause of the new star. He praised Kepler's industry in arguing from sublunary things to celestial phenomena, but disagreed with the form that such argumentation took in this case. He wrote:

> Here let me explain, please, what prevents me from subscribing to this opinion. You know, my Kepler, that the nature of our earth is far different from that of the stars, since the bodies of the latter are simple, similar and most pure, as Aristotle attests. On the other hand, our earth is not simple but dissimilar and impure, and various bodies of diverse substance and temperament are mixed with it. Since, however, in order to produce an action two principles are necessarily required, that is, an active and a passive principle (for everything that acts does so not on itself but on something else), there is therefore a place given here on our earth where various humours and juices occur, and the heat of the earth is able to exercise its action on them by attenuating and resolving them into vapours and exhalations. But what will you give in the stars? Some kind of soluble substance? Will you perhaps say that the stars are imbued or pregnant with some humour or juice, which is then resolved into vapours? But then they are not pure and simple bodies.

ex qualibet redundanti materia fieri commodissime possit; propterea sudorem faeminae et canis convertit in pulices capitis, in pediculos, rorem in bruchos, erucas, linum in anguillas, uliginem in ranas, aquam in pisces, terram in plantas, cadaver in vermes, stercus in scarabaeos, et infinita nova ac insolentia, aeris halitus in διατάττοντας, aethereae regionis in cometas tandem et stellas. Videas ubivis existere uliginosum quippiam, quod rationem continet seminariam, efficitque specierum varietatem, ut non ex cuiusvis arboris foliis quaevis eruca nascatur, sed ex singulis fere singulae. Ille vero totius mundi spiritus hoc praestare videtur, ut omnia invicem ordinentur, accedant novae creaturae instrumenta corporis convenientia. Si nihil viveret, ipse totam materiam vivificaret, ut Terra, si nihil moveretur, ipsa omnia ad se attraheret, alia propius aliis.' On Cornelius Gemma (1535–1578), see Hirai 2008.

[22] Such a comparison is made in Boner 2008. However, while Boner's analysis of Kepler is good, some passages from Gilbert are misinterpreted.

[23] Kepler 1937–, I, 269: 'Haec nempe est illa, quae quoties invenit superfluam aliquam materiam; convertit eam in animalculum tale, quod rerum naturae serviat, seu juvandae seu exonerandae'.

Then to whose profit (*cui bono*) do this humour (or another analogous material), and the exhalations into which it is resolved, serve? Perhaps so that the natural faculty of the rest of the ether may have a way to purge itself of its own excretions? In our globe, the function of these exhalations is different, for their own specific purpose is to be turned into rain, snow, dew and winds, by which the earth is made moist and fruitful, and the air is cleansed. All these things then serve generation, so that plants and fruits germinate for the nutriment and sustenance of all living things. No one denies this use to be superfluous to the globes of the stars, unless one with Giordano Bruno were to construct as many worlds as there are globes of the world.

And to put it briefly, the production of excrements usually accompanies either generation or nutrition, or an action serving these, namely digestion. Thus where nature establishes no generation, or nutrition, there it produces no excrement. And therefore unless it is demonstrated either that stars generate, or are nourished, or are inhabited by animals which are in need of nutrition, it is laboured in vain concerning the birth or purification of excrements rising from that place.[24]

Brengger seems to have been reluctant to abandon the Aristotelian distinction between the impurity of the sublunary realm and the purity of the celestial one.[25] The attribution of some passive fluid material capable of being acted upon and resolved into vapours contradicted this axiom of natural philosophy, and so Kepler's theories of the generation of the new star were invalid. On top of this, Brengger suggested that Kepler's arguments lead to the untenable conclusion that there are living creatures in the stars, and indeed an infinite number of worlds. His use of pluralism and ET life as a *reductio ad absurdum* argument is a reminder of just how left-field

[24] Brengger to Kepler (1 Sep 1607), no. 441 in Kepler 1937–, XVI, 38–40: 'Hic quaeso patiaris ut exponam quid me impediat quo minus huic sententiae subscribam. Nosti mi Keplere longe aliam esse nostrae terrae quam stellarum rationem, siquidem harum corpora sunt simplicia, similaria, et teste Philosopho purissima: nostra autem terra non simplex sed dissimilaris et impura, cui varia diversarum substantiarum et temperamentorum corpora sunt permixta. Cum autem ad actionem perficiendam necessario requirantur duo, Agens et Patiens: (quicquid enim agit non in se agit sed in aliud) hinc actioni et passioni locus quidem datur in nostra Tellure ubi varii occurrunt humores et succi, in quos calor terrae actionem exerere potest eos attenuando et in vapores ac halitus resolvendo. Quid vero in stellis dabis? Quam substantiam resolubilem? An et illas humore aliquo aut succo perfusas seu praegnantes dices, qui in vapores resolvatur? At sic non sunt pura et simplicia corpora. Deinde cui bono inserviunt tum humor ille, vel analoga materia alia, tum exhalationes in quas resolvitur? An ut reliqui aetheris facultas naturalis habeat quo subinde repurget inquinamenta? In nostro globo alia eorum est utilitas: nam exhalationes suum habent finem, ut nimirum vertantur in pluvias, nivem, rorem, ventos, quibus humectetur et foecundetur terra, ac mundetur aer. Ista omnia deinde inserviunt generationi, ut germinent herbae et fruges, in nutrimentum et sustentationem omnium animalium. Quem usum stellarum globis supervacaneum esse, nemo negat, nisi qui cum Jordano Bruno Nolano tot mundos statuit, quot sunt globi mundani. Et ut summatim dicam, excrementorum proventus comitari solet vel generationem vel nutritionem, vel actionem illis famulantem, puta coctionem. Itaque ubi natura nullam instituit generationem, aut nutritionem, ibi nulla producit excrementa: et proinde nisi demonstretur vel stellas generare, vel nutriri, vel inhabitari ab animalibus quae nutritione egent, frustra laboratur de excrementorum inde consurgentium exortu, aut repurgatione.' For Aristotle's theory of the incorruptibility of the heavens, see *On the Heavens*, I, 3.

[25] A more promising connection between the two realms, in Brengger's opinion, could be found in Gilbert's magnetic philosophy, which he thought 'pointed the way to the inner sanctuaries of celestial philosophy (*ad penitiora Philosophiae coelestium adyta viam monstrare mihi videntur*)': Kepler 1937–, XVI, 40.

these ideas were at the beginning of the seventeenth century. Brengger invoked the *cui bono* question to argue not for pluralism, but against the extension of natural terrestrial processes into the celestial realm. The existence of living creatures on other worlds is absurd, and so the action of generation or exhalation in the heavens must be ruled out on the basis of its superfluity.[26]

In his reply, however, Kepler was rather happy to accept the conclusion which to Brengger seemed so ludicrous. He wrote:

> You consider the globes of the stars to be the most pure and simple. They seem to me to be similar to our earth. You, being a philosopher, quote the Philosopher [i.e., Aristotle]: if he were asked, he would point to experience. But experience is wanting, because no one has been there, therefore it does not deny or affirm. But I myself argue with probability from similarity, and by induction, as you do, from the moon, which has many phenomena similar to the terrestrial ones. I therefore grant to the stars both humour and also regions which are rained upon by the exhalation of that humour, and living creatures, for which this cycle turns into something useful. Not only unfortunate Bruno, burned on the coals in Rome (*Prunus prunis tostus Romae*), but also my Brahe was of this opinion, that there are inhabitants of the stars (*stellis incolae*). I follow this opinion so much more willingly because with Aristarchus I affirm the earth to move as the planets do.[27]

Here we see Kepler using analogy and induction to take what he has established about the moon and extend it to the stellar region more broadly. There is an interesting distinction which presents itself here between the cosmologies of Kepler and Gilbert on the issue of the vacuum. Gilbert maintained a strict distinction between, on the one hand, the globes and their effluvia and, on the other, the void deprived of all substance and activity (see Sect. 3.3). Kepler, at least at this stage in his life, saw the heavens as composed of a certain ethereal air, which was largely continuous with the air of the earth, the latter albeit mixed with vapours. The interplanetary and interstellar regions are thereby characterised not by privation but by both substance and activity, allowing for the production of celestial novelties via the action of a pervasive generative faculty analogous to that of the earth. Kepler of course did not feel that by placing inhabitants in the stellar region he was committing himself to an infinite universe, but rather a finite world with more than one centre of life. In this sense he was happy to place himself alongside Bruno on the issue of ET life.

---

[26]Another figure who took a similar line of argument with Kepler was Helisaeus Roeslin (1545–1616), who believed the pluralist consequences of Copernicanism to be sufficient grounds to reject it. He also criticised the homogeneity of the celestial and terrestrial realms on the grounds that it would break down the astrological causal chain. See Granada 2011b.

[27]Kepler to Brengger (30 Nov 1607), no. 463 in Kepler 1937–, XVI, 84–92: 'Stellarum globos putas purissimos simplicissimosque, mihi videntur esse similes nostrae Telluris. Tu philosophum allegas philosophus: si rogaretur experientiam diceret. At tacet experientia, cum nemo ibi fuerit, igitur nec negat, nec affirmat. Ipse vero argumentor probabiliter a similitudine, et ut tu inductione a Luna, quae multa habet similia terrestribus. Itaque et humorem stellis tribuo et regiones, quae ab exhalatione humoris compluantur et creaturas viventes, quibus id utilitati cedit. Nec enim solus infelix ille Prunus prunis tostus Romae, sed etiam Braheus meus in hanc concessit sententiam, esse stellis incolas. Id ego tanto libentius sequor, quod, ut planetas, sic Tellurem etiam ferri affirmo cum Aristarcho.'

Of course, Brengger could have known all this from *De stella nova* itself, in which Kepler did indeed assert the existence of life amongst the stars. Kepler did not, however, make the assertion in connection to the natural cause of the new star or the alterability of the heavens. The topic of ET life instead came up when he at last tackled the issue of the astrological significance of the nova. The position of the new star, so close to the conjunction of Jupiter, Saturn and Mars, seemed to suggest the hand of providence above the natural method of its creation. There would of course be some astrological effect, because nothing was deemed to happen in the heavens which was not occultly perceived by the faculty of the earth and those of its inhabitants. But in determining its *particular* significance, Kepler first had to establish *to whom* it was significant:

> Therefore, having assumed that this star was set alight by the decision either of God himself or a rational creature, if I am asked to what end I think this was made and whether the phenomena signified by it are to be referred to what we as human beings have before our eyes, first I think that not only the individual nations, but the whole earth is too small, so that all our thoughts that arise about the genuine meaning of the star dwelling in the highest part of the ether are squandered in the attempt to embrace it. And the reason is that the size of the world is great, and Tycho Brahe did not reject as absurd that opinion of certain ancient philosophers who stated that the other spheres, which are so immense, also had their own inhabitants, certainly not human beings, but other creatures. And if these exist, they would assuredly suit the beauty of the world, and they not only feel deeply such a great effect of the stars, but they are also embraced by the providence of the supreme custodian. Nor will it be absurd if that sign, sent from that highest observatory of the fixed stars as much for them as for us men dwelling on earth, is perhaps more appropriate for their understanding than for ours. I ask therefore not for which *race*, but for which *globe altogether* this star can more aptly be believed to have been lit.[28]

Of course, Kepler was in no position to investigate what exactly the effect of the new star on these other creatures would be, so he turned his attention to its terrestrial impact. Yet this is evidence that he considered celestial influence to be operative at locations other than the earth, in both a physical and a more occult sense, for the benefit of the non-human inhabitants therein. This, significantly, was the first time Kepler mentioned the possibility of ET life existing somewhere other than the moon, and it is mentioned in reference to astrology. There is a good reason for this. While one could see, for example, the mottled surface of the moon, observational

[28] Kepler 1937–, I, 339: 'Igitur posito quod certo consilio seu Dei ipsius seu creaturae rationalis incensa fuerit haec stella, si ex me quaeretur, quem ad finem hoc factum putem; et utrum eius significata ad ea. pertrahenda sint, quae sub manibus habemus homines? Primum ego non gentes tantum singulas, sed totum adeo Telluris globum nimis exilem puto, ut in eius complexum omnes cogitationes nostrae quae oriuntur super genuina significatione, sideris in altissimo aethere versantis, effundantur. Magna namque Mundi amplitudo est; nec absurda Tychoni Braheo visa est illa veterum quorundam Philosophorum opinio, statuentium caeteris quoque globis, qui vastissimi sunt, suos esse incolas, non equidem homines, at creaturas alias; quae si sunt, ad mundi ornatum utique pertinebunt, neque tantum siderum effectus persentiscent; sed etiam providentia supremi custodis comprehendentur; nec absurdum erit, aeque ipsis ac nobis in tellure versantibus hominibus, ex illa altissima specula fixarum Sphaerae signa mitti, magis forsan ipsorum appropriata captui, quam nostro. Quaero ergo non cui genti sed omnino cui globo potius credendum sit accensum esse sidus hoc.' Emphasis added.

evidence linking the other planets qualitatively to the earth was hard to come by. This is demonstrated by Kepler's belief that the planets shone with an intrinsic light. The other possible line of argument would be if one could prove the existence of certain other celestial phenomena which could not be sufficiently explained on the basis of terrestrial utility. In this way, astrology came to the forefront of *cui-bono* arguments in favour of ET life.

4.4 The New Astronomy: Influence of, and Departures from, Gilbert

Kepler's famous *Astronomia nova* ('New Astronomy') was not published until 1609, but a draft was complete in 1605, at the time when the excitement about the new star was in full swing.[29] While it does not contain any discussion of ET life, it does provide an opportunity to demonstrate just how greatly his celestial physics was influenced by Gilbert's *De magnete*. Later in his *Epitome Astronomiae Copernicanae* ('Epitome of Copernican Astronomy', 1617–1621), Kepler would declare that he built his whole astronomy upon Copernicus' hypotheses, Tycho's observations, 'and lastly upon the Englishman, William Gilbert's philosophy of magnetism' (Kepler 1937–, VII, 254, 1952, 850).[30] The cosmologies of the two men share much in common, but when Gilbert the physician and experimentalist comes up against Kepler the geometer and metaphysician, certain crucial differences arise. Kepler acquired a copy of *De magnete* in 1602, and shortly after this his friend Hans Georg Herwart von Hohenburg (1553–1622) wrote to ask his opinion on the work, particularly seeing as how Gilbert, in Herwart's view, seemingly denied the annual motion of the earth.[31] Kepler soon realised the profound implications that the magnetic philosophy could have for his endeavours in astronomy and celestial physics in particular. In the *Astronomia nova*, as we shall see, he truly ushered in a new astronomy, similar to Gilbert's in its meld of the mechanistic and animistic, but far more advanced and ultimately influential.

In the dedicatory letter to the work, Kepler explained what could be considered a maxim of his entire philosophy:

> Indeed, when I was but indifferently well versed in this theatre of Nature, I formed the opinion, with practice (*usus*) as my teacher, that, just as one human being does not greatly differ from another, neither does one star differ much from another, nor one opponent from another, and hence, no account is to be received easily that says something unusual about a single individual of the same kind.[32]

---

[29] Kepler, *Astronomia nova* (1609), in Kepler 1937–, III. Translations from Kepler 1992.

[30] Bialas sees the three main influences for Kepler's natural philosophy, especially in terms of metaphysics, as Cusa, Scaliger and Gilbert. See Bialas 2009, 29–30.

[31] On Herwart von Hohenburg, see Boner 2014.

[32] Kepler 1937–, III, 9: 'Quippe cum essem in hoc Naturae theatro mediocriter versatus: illud me, usu Magistro, didicisse persuadebar, non multum distare, ut hominem ab homine, sic neque stel-

A similar mode of thought runs through Gilbert, who, while allowing for differences between individual bodies, argued for the basic similarity of all the primary globes of the universe, including the earth with its magnetic astral virtue. But whereas Gilbert did not acknowledge a position in regard to the annual motion of the earth (see Sect. 3.4), Kepler spent the first part of the *Astronomia nova* explaining why, on the basis of physical reasoning, the Tychonic system should in fact lead to the Copernican one.[33]

A major objection to the displacement of the earth from the centre of the universe concerned the concept of gravity. Kepler, like Gilbert, believed that the common theory was in error. His own theory described gravity as 'a mutual corporeal disposition among kindred bodies to unite or join together' (Kepler 1937–, III, 25, 1992, 55). This is in line with the theory of gravity we see in Plutarch and Copernicus, with the key term being 'kindred' (*cognatus*). The earth and the moon are two such kindred bodies, embracing one another within their spheres of influence, resulting in phenomena such as the tides. But Kepler did not see gravity as having any application to the issue of orbits. He maintained, therefore, that the moon and the earth would come together if they were not held apart by an animate force or something similar (Kepler 1937–, III, 25). He further maintained that all corporeal substances have a tendency to rest; that is, they possess an inherent inertia, which in his understanding meant a resistance to motion relative to a body's size and density.

This understanding affected Kepler's theory of planetary orbits. He identified the cause of planetary motion as an immaterial species proceeding from the sun, which weakens according to a law of diffusion as it travels *sine tempore* out to the sphere of the fixed stars (Kepler 1937–, III, 34). The weakening grasp of the sun's power, coupled with the respective bulk and inertia of the celestial bodies themselves, accounts for the slower orbital speeds of the outer planets. The major discovery of the *Astronomia nova* was that these orbits were not circular, but elliptical. He saw in magnetism the potential to provide a physical explanation for the elliptical movements of the planets, which entailed varying speeds and distances from the sun. He considered the magnetic faculty as another example of the sort of mutual corporeal disposition which resulted in gravity. He was satisfied, as were many others at the time, with Gilbert's proofs that the earth was a magnet, and was equally happy that Gilbert had proven the same for the other planets (although Gilbert had in fact said that they have their *own* powers, analogous to magnetism in the earth). Kepler suggested that there was a motive power responsible for making a planet's body reciprocate along a straight line extended towards the sun, and that this was in fact a magnetic corporeal faculty belonging to the planetary bodies themselves (Kepler 1937–, III, 34–35, 348–364).

This scheme prompted Kepler to declare that 'every detail of the celestial motions is caused and regulated by faculties of a purely corporeal nature, that is, magnetic' (Kepler 1937–, III, 35, 1992, 68). The exception remained, however, that the rota-

---

lam a stella, hostem ab hoste: quare non facile recipiendum sermonem, qui de gentis eiusdem individuo uno temere aliquid insolitum sparsisset.' Kepler 1992, 33.

[33] On the importance of the physical reality of heliocentrism for Kepler, see Gingerich 1973.

tion of the sun required a vital faculty, or the action of a soul. Just as in Gilbert, magnetism could explain a fixed axis of rotation, but the actual movement of rotation required an impetus, either external or internal. Nevertheless, Kepler was rightly pleased with his new astronomy. He could now declare arguments based on the common notion of planetary intelligences or angels to be irrelevant to considerations of the natural operation of planetary movement. The argument was now 'about powers not endowed with a will to choose how to vary their action', he stated, 'and about minds which are not in the least separate, since they are yoked and bound to the celestial bodies which they are to bear'.[34]

Kepler's description of the motive power of the sun made its operation analogous to that of light in a way almost identical to that described by Gilbert. These two immaterial species of the sun travel instantaneously, not existing in the intermediate space, manifesting only when encountering some suitable material. When this motive virtue, which Kepler described as the 'the first actuality of every motion in the universe' (*actus primus omnis motus mundani*), comes into contact with a mobile body, such as a planet, it 'actualises' (*in actum elicit*) the planet's own proper power, similar to what we saw in Gilbert. It also mixes with the body's own inertial disposition, ultimately resulting in the observed motion of the planet. Kepler compared this to the way in which colour mixes its own species with light to create a 'third entity'. Here again is an attempt to explain diverse phenomena as results of analogous processes. Kepler and Gilbert both believed in a highly animistic universe, but one which endured with a quasi-mechanistic operation, and which was ultimately reducible to a few fundamental laws.[35] The application of terrestrial physics to the question of celestial motion is what aligns them in the common narrative of the history of science. One important difference is that while Gilbert was sceptical of the ability of astronomy to map or predict the resulting motions, Kepler wanted or in fact required his physical hypotheses to produce definite distances and equations (Kepler 1937–, III, 364).

4.5 Alien Worlds Through the Telescope

The Italian philosopher Galileo Galilei (1564–1642) is credited as a harbinger of the Scientific Revolution for, among other things, his application of mathematics to physics, and of physics to astronomy. He had a major impact on Kepler's astronomy, not through his laws of motion—which included a more accurate theory of inertia—but through his telescopic observations. The enhanced view of the universe

---

[34] Kepler 1937–, III, 98: 'Disputamus enim de rebus naturalibus dignitatis longe inferioris, de virtutibus nullo arbitrio ad variandam actionem suam usis, de mentibus minime sane separatis cum sint conjunctae et alligatae corporibus coelestibus vehendis.' Kepler 1992, 170.

[35] That is to say, just because the earth and other heavenly bodies may be alive does not mean they act in an inconsistent way. Once you understand the *telos* of a living object you can formulate rules for its behaviour under certain conditions.

provided by the telescope was one of the biggest spurs to scientific, philosophical and cultural interest in ET life. At the same time, it presented new challenges to the theoretical foundations of astrology.[36] The results of Galileo's observations were published in his *Sidereus nuncius* ('Sidereal Messenger', 1610). If these results were to be believed, it would necessitate radical revisions of the size, substance and contents of the celestial region. When Kepler read it he immediately accepted the veracity of Galileo's observations, and was thus forced to rethink several key features of his astronomy. He discussed these issues in his hasty reply to Galileo, which was soon after published as the *Dissertatio cum nuncio sidereo* ('Conversation with the Sidereal Messenger', 1610).[37] For example, he admitted that he was wrong about the land/sea division on the moon. He had previously thought that the dark areas were land, with the brighter areas being reflective oceans. Galileo's arguments proved to him that in fact the opposite was the case (Kepler 1937–, IV, 298). Another adjustment came from the fact that there seemed to be no indistinctness in the image as a result of the passage of the light through the ether. 'Hence we must virtually concede, it seems,' concluded Kepler, 'that that whole immense space is a vacuum.'[38]

The key word here, *pēne*, translated by Rosen as 'virtually', could also be read as 'nearly' or 'almost', and it's important to realise that Kepler was not in fact ready to believe that the space between the planets was a void.[39] It would have ruled out the plenist philosophy which underpinned his theories, as discussed above, on the production of celestial novelties. It would also have conflicted with his theory of the universe as an image of the Trinity, with the sun as the Father, the sphere of the fixed stars as the Son, and the space in between as the Holy Spirit. This was a theory with a deep and long-lasting significance for Kepler, and its influence can be perceived in his constant poetic praise of the sun, as well as his engagement with cosmological issues such as the vacuum and the infinity of the universe.

The size of the universe was indeed another issue to be brought into question by telescopic observations. Kepler was led to the topic in his *Dissertatio* first of all through Galileo's conclusions about the circular appearance of the planets in contrast to the sparkling points of the stars. The issue was finally settled for Kepler: the stars generated their own light, but the planets were wholly illuminated from without. He even conceded to a point of Brunian philosophy: that the stars were suns and the planets were moons or earths. Depriving the planets of an intrinsic light did not in fact have major consequences for Kepler's astrology. The planets' role in celestial influence was almost wholly contained in the 'geometrical cause', and we have seen

[36] On Galileo's involvement with astrology, see Rutkin 2005.

[37] Kepler, *Dissertatio cum nuncio sidereo* (1610), in Kepler 1937–, IV, 283–311. Translations from Kepler 1965.

[38] Kepler 1937–, IV, 294: '… ut pene concedendum videatur, totum illud immensum spacium vacuum esse'. Kepler 1965, 19.

[39] He was not alone in this. Gilbert was almost unique in arguing for an interplanetary void at the beginning of the seventeenth century, while Kepler's commitment to a celestial air or resistanceless ether would remain commonplace well into the eighteenth. On Newton's transition from a material to an immaterial ether, see Grant 1981, 247.

how the individual natures were preserved through the interaction of light with the planetary essence, represented by colour. What this discovery most likely did do was bolster Kepler's confidence in extending, by induction, terrestrial analogies beyond the moon to the other planets.

Yet Kepler was not prepared to consider an infinite universe. He believed that the appearance of an incredible crowding of stars through the telescope demonstrated that, in fact, those bodies were close together, and that 'where we mortals dwell, in the company of the sun and the planets', provided a unique perspective and was indeed the very heart of the universe (Kepler 1937–, IV, 302–3, 1965, 34). Nor would Kepler concede that there may be planets or moons orbiting the fixed stars. This last option had been what Kepler had feared, and his friend Johannes Matthäeus Wackher von Wackenfels (1550–1619) had hoped, upon hearing the news that Galileo had discovered four new planets. Kepler dismissed Bruno's reasoning that the fiery worlds of the suns required the company of the watery earths, so that both could be sustained and in turn sustain life.[40] As the four new planets were in fact moons orbiting Jupiter, rather than planets orbiting one of the fixed stars, Kepler felt at ease once more in his belief in a finite universe. 'Hence this will remain an open question', Kepler concluded, 'until this phenomenon too is detected by someone equipped for marvellously refined observations. At any rate, this is what your success threatens us with, in the judgment of certain persons.'[41] Rosen pointed out that this passage was not part of the original letter reply to Galileo, but was added to the published version, possibly in an attempt to appease someone such as the Brunian Wackher (Kepler 1965, 137, n. 340).

It was for this friend Wackher that Kepler professed to have recently written a lunar geography (Kepler 1937–, IV, 297–98). This, we assume, was added to his university disputation to form the *Somnium*. In it he had worked on the assumption, now firm in his mind, that there were indeed living beings on the moon. In the *Dissertatio* we see how this assumption asserts a bias over Kepler's interpretation of observational evidence. On the subject of the appearance of large, circular cavities, Kepler asked whether it was the work of nature or a trained hand (Kepler 1937–, IV, 299). The appearance of geometrical shapes in nature seemed to him always to suggest the work of a mind. He also saw it as highly probable that there was a lunar atmosphere. The evidence for this was thin. He asserted that at full moon the spots on the moon are hidden somewhat and the sun's rays more brightly reflected by dense air. He also states that Maestlin had seen rain on the moon. But what seems to have really convinced him is the assumption that the lunar inhabitants would not be able to endure the intense heat of the sun in the middle of the day (Kepler 1937–, IV, 299). The 'joke' about living creatures on the moon had now become an accepted premise for the argument that lunar conditions must be conducive to life.[42]

[40] See Granada 2009.

[41] Kepler 1937–, IV, 305: 'Hoc igitur in incerto manebit, quoad aliquis subtilitate observandi mira instructus, et hoc detexerit: quod quidem hic successus tuus, iudicio quorundam nobis minatur.' Kepler 1965, 39.

[42] See the analysis by Dick 1982, 178: 'In proportion to the extent that Kepler saw an Earthlike

The interpretation of celestial phenomena through the lens of a pluralist worldview continued when Kepler came to discuss Galileo's discovery of four moons orbiting Jupiter. This discovery could have caused more problems for Kepler's cosmology. First of all, the orbits of the moons could disrupt his system which spaced out the planets according to the Platonic solids. Luckily the orbits could easily be accommodated within the 'sphere' of Jupiter's own orbit, and Kepler hoped they might even help resolve a discrepancy between the actual distance separating the orbits of Mars and Jupiter, and that predicted by his theory (Kepler 1937–, IV, 310).[43] The other problem was in the realm of astrology. Didn't these new planets discredit the predictions of astrologers, who were never aware of them or their potential influence? Thankfully Kepler knew that the planets do not influence us directly, but only through the perception of aspects by sublunary souls. The moons were not visible without the assistance of the telescope, and more pertinently their orbits close to Jupiter's globe did not create an intelligible difference to the formation of aspects with the other celestial bodies (Kepler 1937–, IV, 306).

For whose sake, the question then became, were there these four small bodies revolving around Jupiter at different distances with different periods? The answer was that there must be creatures on Jupiter to observe this 'wonderfully varied display' (Kepler 1937–, IV, 306, 1965, 40). Directly after determining that the new moons had no role to play in terrestrial astrology, Kepler concluded that they must therefore have been ordained for the Jovian beings. The *Sidereus nuncius* led Kepler to add a new category of celestial object, *circulatores* (satellites), to his cosmos:

> The conclusion is quite clear. Our moon exists for us on the earth, not for the other globes. Those four little moons exist for Jupiter, not for us. Each planet in turn, together with its occupants, is served by its own satellites. From this line of reasoning we deduce with the highest degree of probability that Jupiter is inhabited.[44]

Just like that, the conjecture that Jupiter was inhabited became an assertion that *each* planet had its own inhabitants. Not only that, he also concluded that each must have moons ministering to it, even though none of these had as yet been observed. There is an interesting parallel here to Benedetti's theories (see Sect. 2.5), except that whereas he had seen the planets as moons and posited the existence of other earths to accompany them, Kepler sees the planets as earths and grants each of them moons. The belief in ET life was influencing the interpretation of celestial phenomena and also informing predictions about the reality of the cosmos beyond the limits of observations.

---

moon as a prediction of the Copernican theory, that theory may have affected his interpretation of the evidence for a lunar atmosphere. Such arguments of interpretation, enmeshed in metaphysical predispositions, rendered it virtually certain that observation would provide no speedy solution to the problem of a world in the moon, much less in the planets.'

[43] On Kepler's continuing commitment to this theory, see Field 1988, 73–95.

[44] Kepler 1937–, IV, 307: 'Plane igitur sic est, quod nobis est in Tellure nostra Luna, hoc non est globis caeteris; et quod Iovi sunt illae quatuor lunulae, id non sunt nobis: et vicissim singulis planetarum globis, eorumque incolis, sui serviunt circulatores. Ex qua consideratione, de incolis Iovialibus summa probabilitate concludimus.' Kepler 1965, 42.

As with the nova of 1604, talk of ET life went hand in hand with considerations of celestial influence. Just as the influence of the new star would be more significantly felt by an audience somewhere other than on the earth, the exclusion of the new moons from terrestrial astrology led to the conclusion that they were intended to influence the non-human inhabitants of Jupiter. The benign influence provided by orbiting lunar servants became an indication of life. Then, by a process of induction, it became a necessary condition thereof. This of course presents a seeming inconsistency in Kepler's pluralism. If moons or satellites were intended for the benefit of the inhabitants of their respective planet, where do Kepler's lunar inhabitants fit in? This is indicative of the different methodologies which Kepler applies to the earth's moon on the one hand, and the larger solar system on the other. But perhaps there is no inconsistency. The same globes could be both sources of influence and dwelling places for ET life. Though Kepler never mentioned the possibility of life on the Jovian moons as well as the globe of Jupiter itself, there is no reason to think he would have denied the possibility. Kepler's cosmology thus shares a certain commonality with Bruno's vision of binary planetary opposites. The difference is one of complexity and indicates that Kepler preferred a finite cosmos defined by diversity rather than the repetitious uniformity implied by an infinite universe.

The existence of life on the other planets raised for Kepler the religious and philosophical question of its implications for the importance of mankind in the universe. This is a question which has dominated pluralist philosophy ever since, and its relevance is still felt today.[45] How can we be sure, Kepler asked, that we are the most noble of all creatures? How can we be sure that we occupy the central role in God's providence? We have seen already in the *Mysterium cosmographicum* that part of Kepler's answer was that the earth, occupying a middle position between the planets and the two classes of solids, represented the most prestigious location in the cosmos. It had a temperate climate and afforded a view of all the planets. At the same time, Kepler felt that the four moons given to the Jovians served as compensation for their distance from the sun and their inability to see the inner planets. Kepler even went further, suggesting that the moons were established upon a blueprint of geometric solids just as the primary planets were:

> Let the Jovian creatures, therefore, have something with which to console themselves. Let them even have, if it seems right, their own four planets arranged in conformity with a group of three rhombic solids. Of these, one is the cube (a quasi-rhombic); the second is cuboctahedral; the third is icosidodecahedral; with 6, 12, and 30 quadrilateral faces, respectively. Let the Jovians, I repeat, have their own planets. We humans who inhabit the earth can with good reason (in my view) feel proud of the pre-eminent lodging place of our bodies, and we should be grateful to God the creator.[46]

---

[45] This is naturally a very broad topic. Some good introductions are Dick 1996; Losch 2016; Mix 2016.

[46] Kepler 1937–, IV, 309: 'Habeant igitur creaturae Ioviae quo se oblectent; sint illis etiam, si placet, quatuor sui planetae dispositi ad normam classis trium rhombicorum corporum; quorum unum (quasi rhombicum) cubus ipse est, secundum Cuboctaedricum, tertium Icosidodecaedricum, sex, duodecim, triginta planorum quadrilaterorum: habeant inquam illi sua: nos Homines Terricolae non utique frustra (me doctore) de praestantissima nostrorum corporum habitatione gloriari pos-

Kepler never developed this theory beyond its brief mention in the *Dissertatio*. However, it was one of the axioms of his philosophy that the attributes of geometrical shapes have consequences in the natural world, both astrological and otherwise. We can therefore imagine that he saw these solids as dictating more than just the Jovian lunar distances.

In the *Dissertatio*, Kepler urged Galileo to establish the astronomy of Jupiter for the benefit of mankind's future space explorers, while he himself established that of the moon. From the above passage, we see that Kepler could not resist the temptation to lend his own talents to Jovian astronomy. He continued to do so in his *Dioptrice* (1611), his second work on optics which now included an investigation into the science of the telescope and its observations. In the space of a few paragraphs he described the appearance of the heavens from Jupiter, just as he had done for the moon:

> And certainly in these places, this sun of ours, common focus both of this earthly world and that Jovian world, which we estimate to be of the size of thirty minutes, fills up barely six or seven minutes, and having measured the zodiac in the space of twelve of our years is found to be in the same fixed stars again. Therefore, those creatures living in that globe of Jupiter, while they contemplate those brief courses of the four moons through the fixed stars, while they look at them and at the sun rising and setting every day, would swear by the Jupiter Stone (indeed I have recently returned from these regions) that their own globe of Jupiter rests immobile in one place (while in truth the bodies that rest are the fixed stars and the sun), just as their four moons revolve with a multiple variety of motions around their home.[47]

The issue of perspective again becomes a way for Kepler to counter arguments against the motion of the earth. He paints the reader a picture of the sky as seen from Jupiter, with the sun being pitiably small, but with four moons continually tracing their courses against the backdrop of the fixed stars. This practice of considering the apparent motions of the celestial bodies from a point in the universe other than the earth would take on a new significance in Kepler's cosmological masterpiece, the *Harmonice mundi*.

---

sumus, Deoque conditori grates debemus.' Kepler 1965, 46. Kepler would later try and prove this rhombic solid theory. See Field 1988, 79–80, 218–29.

[47] Kepler 1937–, IV, 343–44: 'Atque illis quidem locis Sol hic noster, communis et huius terrestris, et illius Iovialis mundi focus, quem nos tricenum plurimum minutorum esse censemus, vix sena aut septena minuta implet; interimque duodecim nostratium annorum spacio Zodiacum emensus apud easdem rursum fixas deprehenditur. Itaque quae in illo Iovis globo degunt creaturae, dum illa quatuor Lunarum brevissima per fixas curricula contemplantur, dum quotidie orientes occidentesque et ipsas et Solem aspiciunt, Iovem lapidem jurarent (nuper enim ex illis regionibus reversus adsum) suum illum Iovis globum quiescere uno loco immobilem, Fixas vero et Solem quae corpora revera quiescunt, non minus quam illas suas quatuor Lunas multiplici motuum varietate circa suum illud domicilium converti.'

4.6 Harmony and Astrology

Johannes Kepler's *Harmonices mundi libri V* (1619) was the culmination of a research project which had occupied him for two decades.[48] We have seen how in his *De fundamentis* he had suggested a third astrological cause based on the ratio between the orbital speeds of the planets. The main achievement of the *Harmonice mundi* was Kepler's discovery that the extreme velocities of the planets fit very closely to a musical scale, but not from the perspective of the earth where he had first looked. The scale only fit when the orbits were calculated from the vantage point of the sun. This discovery forms the climax of the work in Book 5, but the work as a whole is a much larger investigation into the origins, causes and attributes of harmony, and the application of the science of harmonics to human, natural, and celestial philosophy. It will be worthwhile to briefly track Kepler's development of a harmonic philosophy, especially through the second half of the *Libri V*, as the scale of his attempt to redefine astrology within a non-geocentric framework has not been properly grasped.[49] This will also provide a better understanding of just how relevant the question of astrology is to ET life in Kepler's harmonic heliocentric cosmos.

The first two books of the *Harmonice* deal with the construction of the regular polygons and polyhedra, and the congruence thereof. Kepler thus made it clear that the science of harmony belonged firmly within the science of geometry, rather than arithmetic or numerology. From that starting point he could assess the geometrical qualities of certain figures, their ratios and congruences, and thereupon build a foundation for a new philosophy of harmonic influence. In the introduction to Book 2 he explained this as an axiom of his entire philosophy:

> …the effect these figures have in the realm of Geometry, and in that part of Architechtonics which deals with Archetypes, is as an image of and a prelude to their effects beyond Geometry, beyond things conceived in the mind, namely their effects in things natural and celestial.[50]

One example of this in practice is when Kepler divided congruence into three distinct classes, along with a fourth class containing no congruence. These classes, he declared, will help in Book 4 when it comes to choosing which aspects are efficacious and in what way (Kepler 1937–, VI, 89).

---

[48] Kepler, *Harmonices mundi libri V* (1619), in Kepler 1937–, VI. Translations are from Kepler 1997, with minor adaptations.

[49] There is something of a disconnect between Kepler's supposedly limited astrological theory, as it is understood by Field, Boner and others, and his more traditional and complex astrological practice, as examined by, among others, Greenbaum. The analysis in this section will hopefully contribute to bridging that divide. See Field 1988, 127–41; Greenbaum 2015.

[50] Kepler 1937–, VI, 67: '…cum hic figurarum effectus intra Geometriam, intraque Architectonices partem illam, quae circa Archetypos versatur, sit quaedam velut imago et praeludium Effectuum extra Geometriam, extraque mentis conceptus, in ipsis rebus naturalibus et coelestibus…' Kepler 1997, 97.

Book 3 of the *Harmonice* deals with musical harmony, and is split into two parts: causes and effects. In the first part Kepler explained how the cause of harmonies lies in the ratio of the arcs of a circle cut by the inscription of the regular polygons (Kepler 1937–, VI, 101–120). This was in opposition to the Pythagoreans who, ignoring the evidence of their ears, ascribed the root of harmony to simple numerical ratios. The much larger part of Book 3 deals with the actual effects of audible harmonies. In fact, the structure of the work makes more sense if we take into account the preamble and explanation of order at the beginning of Book 4. There Kepler explained that there could have been six books: three on harmonic theory (shapes, congruence, musical harmony), and three on its manifestations and applications in the natural world (music, aspects, planetary motion). For some reason Kepler decided to merge the middle two books into one.

Accordingly, part way through the third book Kepler explained the philosophical basis of harmonic efficacy. 'Indeed all spirits, souls, and minds are images of God the Creator', he argued, 'if they have been put in command each of their own bodies, to govern, move, increase, preserve, and also particularly to propagate them.'[51] These minds therefore contain the same archetypal geometric patterns that existed coeternally with God and provided the blueprint for all Creation. These patterns are then observed in the functions and operations of those souls. For the same reason, these souls 'rejoice' when they perceive these archetypal patterns, either by bare contemplation, through the interposition of the senses, or by some concealed instinct (Kepler 1937–, VI, 105). The harmonic scale of planetary motions and the effect of the aspects on the soul of the earth are two examples of this, while the general effect of harmony on the human soul is another. All three boil down to one principle: 'Everything is lively while the harmonies persist, and drowsy when they are disrupted.'[52]

One of the most interesting sections of Book 3 is Chapter 15, which contains a theoretical attempt to link certain tones, intervals and harmonies with different emotional states.[53] When considering this subject, Kepler argued that it must be separated from the 'general' effect of harmony stated above (Kepler 1937–, VI, 173). This was clearly an attempt to create a fuller scheme of harmonic influence, one which extends beyond the action of general stimulus and couples specific harmonies with specific effects. It is in fact another instance of Kepler creating analogies between geometric and natural qualities. One example he gave is how God dictated the laws of generation according to the harmonic divisions involved in the construction of the pentagon. There were two types of generation to be considered. For plants, which contain their own seeds as individuals, the laws of generation follow 'the ratio of inexpressible terms which is genuine and perfect in itself'. Sexual

[51] Kepler 1937–, VI, 105: 'Dei vero Creatoris imagines sunt, quotquot Spiritus, Animae, Mentes, suis singulae corporibus sunt praefectae, ut illa gubernarent, moverent, augerent, conservarent, adeoque et propagarent.' Kepler 1997, 146.

[52] Kepler 1937–, VI, 105: 'vivunt omnia, durantibus Harmoniis, torpescunt iisdem disturbatis'. Kepler 1997, 147.

[53] 'Qui modi vel toni, quibus serviant affectibus', Kepler 1937–, VI, 173–79, 1997, 238–46.

reproduction, on the other hand, is represented by 'the combined ratios of pairs of numbers (of which the falling short of one by unity is compensated by the excess of the other)'. 'What is surprising then', asked Kepler, 'if the progeny of the pentagon, the hard third of 4:5 and the soft 5:6, moves minds, which are the images of God, to emotions which are comparable with the business of generation?'[54] Kepler here explained (or imagined) the aphrodisiac effects of these musical intervals based on the geometrical properties from which they are derived.

The subject of Book 4 is the aspects. This is usually considered the astrological section of the *Harmonice*, which has had the unfortunate consequence of distracting the focus of historians of astrology from the highly relevant information provided at the end of Book 3 (and indeed Book 5, as will be discussed below). The linking of specific harmonies to specific moods forms part of the theoretical groundwork for the practical business of aspect astrology. Before discussing this astrology directly, Kepler began Book 4 by delving further into the essence, perception, and effect of harmonies themselves. Regarding the reception of harmonies by the soul, he asked himself by what means they enter. The answer is partly actively, partly passively. The harmonies are embodied in emanations, or species—sound from the motion of the string, light and colour from the rays of the stars. These emanations are active in the way they move (*movere*) our senses. Yet these same emanations are also passive in the way they are felt (*sentiuntur*), remembered (*memorantur*), and compared (*comparantur*) (Kepler 1937–, VI, 213). Now the harmonies themselves would have no power, and indeed not exist, if there was not a soul to perceive them and compare them to the pure archetypes which exist within itself.[55] The soul has the knowledge of the harmonies through instinct alone, and it is a basic or low faculty which recognizes these proportions in sensible things, belonging not only to men but also to animals, the soul of the earth, and possibly even plants (Kepler 1937–, VI, 226).

The consideration of emanations and instinct is key to understanding the potential complexity of Kepler's astrological system. The inferior faculties of the soul do not experience harmonies instinctually in their purity, but 'along with the wrappings of the emanation (*species*) which is subject to them'.[56] One can therefore differentiate between the effects of the same harmonic proportion coming from different

[54] Kepler 1937–, VI, 175–76: 'Haec cum sit natura huius sectionis, quae ad quinquanguli demonstrationem concurrit; cumque Creator Deus ad illam conformaverit leges generationis; ad genuinam quidem et seipsa sola perfectam proportionem ineffabilium terminorum, rationes plantarum seminarias, quae semen suum in semetipsis habere jussae sunt singulae: adjunctas vero binas Numerorum proportiones (quarum unius deficiens unitas alterius excedente compensetur) conjunctionem maris et foeminae: quid mirum igitur, si etiam soboles quinquanguli Tertia dura seu 4.5. et mollis 5.6. moveat animos, Dei imagines, ad affectus, generationis negocio comparandos?' Kepler 1997, 241.

[55] The pre-existence of the geometrical archetype within the soul of the individual was one of the ways by which Kepler could save the aspects from the criticism of Pico, who saw no reason why an aspect of 60° should be efficacious, while ones of 59° or 61° are not. See Rabin 1997.

[56] Kepler 1937–, VI, 227: 'Quippe ipsae etiam Ideae Harmoniarum, quas hae inferiores Animae facultates secum habent intus, non plane purae ipsis sunt instinctae, sed cum involucris speciei subjectae, eius scilicet quae cuiusque facultatis sit objectum.' Kepler 1997, 310.

sources. The effect is then a result of soul acting upon itself, not by intention but by natural instinct—an instinct which is not only excited by harmony, but which also responds to the various natures of the different proportions:

> And in fact it [the soul] has from its very origin the ideas both of the harmonies incorporated into sound and of the feelings of the mind which respond to them linked together and, so to speak, conflated into one, so that the idea of harmony is implanted in it only insofar as it delights in it, and is something pleasurable, and insofar as it is bound up with the idea of the associated motion.[57]

Considered together, these passages represent the foundation stones of Kepler's vision for a complete system of harmonic astrology, one in which the base level of the soul's excitement by harmony is augmented by the qualities both of the specific harmony or aspect involved and the particular emanations in which it is embodied.

The main, or at least the largest, participator in terrestrial astrology is of course the soul of the earth itself. Kepler believed, like Gilbert, that this sublunary soul or nature is diffused throughout the whole body of the earth. Unlike Gilbert, however, Kepler believed that it would be primarily present, or rooted, in one particular part of the earth's globe, just as the soul of the human body is located in the heart. From there it projects itself through emanations, just as Gilbert's had by effusions, into the oceans and the air (Kepler 1937–, VI, 237). From that central location it also compares the incidence of celestial rays against its instinctual knowledge of the geometrical archetypes of creation and isolates those that are harmonic. By another natural faculty, at those harmonious times, it more energetically operates its 'workshops', located particularly under mountainous regions, resulting in a greater than usual production of vapours and clouds (Kepler 1937–, VI, 237). Hence we have the basis for an astrological prediction of the weather based on the earth's reaction to celestial aspects. As Kepler poetically put it: 'In brief, the configurations sing the leading part; sublunary Nature dances to the laws of this song.'[58]

Kepler described this faculty of the earth, which moves itself according to its archetypes of proportion, sympathy, and associated functions, as an 'activity'. 'Others have named it *dunamis*,' he wrote, 'but I for preference *energeia*. Because this latter is the essence of souls, that essence being a kind of flowing of the flame'.[59] These two Greek terms, fundamental to Aristotelian metaphysics, translate into

[57] Kepler 1937–, VI, 228: 'habetque iam ab ipso ortu connexas et in unum quasi conflatas Ideas, et Harmoniarum in sonos incorporatarum, et affectionum animi respondentium: ut non aliter ipsi sit implantata Idea Harmoniae, quam quatenus laetificat, estque delectabilium aliquid, et quatenus est Ideae motus conformis implexa'. Kepler 1997, 310.

[58] Kepler 1937–, VI, 239: 'Breviter, configurationes praecinunt, Natura sublunaris saltat ad leges huius cantilenae.' Kepler 1997, 325.

[59] Kepler 1937–, VI, 271: 'quam alii δύναμιν, ego ἐνέργειαν lubentius nominaverim. Est enim animarum essentia haec, est veluti ῥύσις quaedam huius flammae ista'. The published English translation (Kepler 1997, 367) renders it thus: 'Others have named it a "power," but I for preference an "activity." For the former is the essence of souls, but this latter a kind of "flowing" of the flame.' The usual interpretation of the Latin *haec* and *ista* as a correlative pair would imply a reverse order, i.e. *energeia* is the essence and *dunamis* is the 'flowing'. The rest of the passage, however, makes clear that for Kepler the essence of the soul *is* its flowing.

Latin as *potentia* (power) and *actus* (activity) respectively. Kepler used 'activity' here in a slightly different sense to that of Gilbert, but it is perhaps significant that both authors prioritised this term particularly to emphasise the animistic nature of celestial influence. It is this frame of animism that allowed Kepler to take his conclusions about meteorological astrology and apply them to the human soul. The principle is indeed the same. Planetary aspects provide an impetus for people 'to carry through the business which they have in hand'.[60] This impetus, which Kepler compared to a goad or a spur, a stirring oration or a symphony of instruments, works more acutely on masses of people than on a particular individual, which is why an aspect involving Mars may lead groups to go to war. However, more interesting to Kepler than this general effect, relevant to the astrological category of 'revolutions', was the issue of horoscopes or nativities.

According to Kepler, an infant, being sustained entirely by its mother during its development in the womb, has its own soul 'lit' at the moment of birth. At this point the vital faculty takes on an impression of the celestial zodiac, including the position of the planets and of course any aspects between them, and retains this impression for the rest of his or her life. Throughout the course of this life, the soul continually compares the situation of the heavens against its own nativity and is spurred or agitated when similar configurations occur (Kepler 1937–, VI, 278–79). Now we have already seen how different harmonic proportions have different effects, like how those associated with the pentagon move minds to generation. Kepler then gave an example of how the emanations which make up the aspect also determine the effect:

> Interpose here, again from birth, the difference which planets make in qualities. For if the soul is a kind of light, it will also distinguish the red colour of Mars from the whiteness of Jupiter and the leaden colour of Saturn. Thus, it must be admitted that great assistance comes from Mars not only, as before, towards industriousness, but also towards sharpness of talents, which is based on fiery vigour.[61]

So the soul of the infant is not moulded by the influence of the planets so much as it moulds itself according to the aspects present at birth, the qualities of those aspects, and the identifiable essence of the bodies which create them.

The depth of Kepler's personal interest in horoscopes has had a light cast upon it by recent scholarship (Rutkin 2006; Juste 2010; Greenbaum 2015). While he professed not to believe in the reality of the zodiacal signs, house divisions, or the ability, for example, of a horoscope to predict death, his system of harmonic astrology could still provide a natal chart to be judged on the traditional basis of planetary qualities and aspects. He also believed, interestingly, in a form of hereditary astrology. Kepler thought that a mother would be more inclined to give birth on days

[60] Kepler 1937–, VI, 277: 'Hinc igitur habent hoc Animae humanae, quod sub tempus aspectuum coelestium praecipuos capiunt impetus, ad negocia, quae sub manibus habent, peragenda.' Kepler 1997, 373.

[61] Kepler 1937–, VI, 279: 'Hic intersere, rursum ex Genesi, Planetarum discrimina in qualitatibus. Nam si Anima lux quaedam est; discernet etiam Martis ruborem a Jovis candore, Saturnique livore: itaque fatiendum est, magnum auxilium ex Marte, non tantum, ut prius, ad industriam; sed etiam ad acrimoniam ingenii, quae consistit in vi ignea.' Kepler 1997, 376–77.

when the configuration of the heavens most closely resembled what it had been at the time of her own birth, therefore creating an affinity of the nativities of parents with their children (Kepler 1937–, VI, 284). In Book 4 of the *Harmonice* he also delved into medical astrology, giving a theory of critical days, which were an example of the astrological category of election, based on the angle between the true moon and its position in the patient's nativity (Kepler 1937–, VI, 286).

The old adage that Kepler dabbled in astrology only as a professional obligation or source of income has been satisfactorily disproven by more recent scholarship. However, one still reads that Kepler disregarded most of astrology, keeping only the aspects (Rabin 1997, 754; Boner 2013, 33–37). The first part of this argument is of course true if one considers how richly complex and varied astrological culture and practice was at the time. Yet looking at Kepler's astrological and cosmological works we see how important astrology was to his understanding of how and why nature, both sublunary and celestial, operated as it did. And while he may have vocally impugned contemporary astrological practice, at the same time he built a new foundation for terrestrial astrology which could encompass three of the four traditional categories of astrological prediction: revolutions, nativities, and elections. The fourth category, interrogations, which was still quite prevalent in medical astrology, seems to have had no place in Kepler's system. Overall, however, the limitation of terrestrial astrology to the aspects did not in fact place much of a limit to the potential application of Kepler's astrology to a wide range of phenomena. Furthermore, when Book 5 of *Harmonices mundi* is taken into account, one cannot even assert that he limited astrology to the aspects at all, or at least not without significant qualification.

4.7 Astrological Harmonies and Solar Beings

Kepler's theory linking planetary orbits to the Platonic solids gave him the reason why there were no more or less than six planets, and also why they were set distances apart. His aspect astrology, based on perspective and therefore transferrable to any centre, could not provide any teleological information regarding the make-up of the cosmos. Planetary motion was the problem which monopolised much of Kepler's astronomical efforts, and he still needed to know why God had created them to move as they do. In Book 5 of the *Harmonice* he gave us his answer. From the perspective of the sun, the proportions between the apparent minimum and maximum orbital speeds of the planets combine to create a musical scale.[62] Indeed, his calculations for these orbital velocities were so close to the prescribed ratios of the scale that Kepler was convinced that musical harmony was the rationale behind the non-uniform motions of the planets. In the course of their elliptical orbits, the planets would vary their 'pitch', a lot in the case of the highly elliptical Mercurial orbit, as little as one diesis for the nearly circular orbit of Venus. Occasionally, these planetary tones would combine in harmonies—a true 'music of the spheres' that was

[62] See Walker 1967.

inaudible yet powerful nonetheless. This was his crowning achievement, and the reason why the *Harmonice mundi* was the work he was proudest of.[63]

The section of Book 5 most relevant to this discussion is the Epilogue, where Kepler, having concluded his calculations and proofs, asked the question: *cui bono*? Here he took the time to philosophise, conjecture and speculate about the deeper reasons behind this planetary chorus that he had discovered. As the harmony is created not by the true speeds of the planets through the ether, but rather by their apparent motions against the zodiac from the point of view of the sun, Kepler argued that 'the harmony does not ornament the ends'. Rather, there must be some audience, some mind which can compare these motions with each other and appreciate the moments when they combine in harmonic proportion, just as terrestrial souls do with the aspects. Kepler thus concluded that 'since no object is arranged vainly, and without something which is moved by it, those angles seem in fact to presuppose some agency, like our sight, or certainly the sensation of it' (Kepler 1937–, VI, 363, 1997, 492). Just as the planets are honoured and served by their orbiting moons, which not only supplement the light from the sun but also create harmonies by their aspects, so is the sun served by the planets, which repay the life-giving gifts of light and heat by combining their motions into 'the most desirable harmony' (Kepler 1937–, VI, 363–64, 1997, 492).

But what is the nature of this mind, or these minds, in the sun, situated as it is in the very heart of the world? He seemed to hesitate between certain possibilities, ranging (if we can create such distinctions in Kepler) from the naturalistic to the philosophical and theological. At first he described the sun as the governmental edifice of the kingdom of nature, populated by 'chancellors, princes or prefects' (Kepler 1937–, VI, 364, 1997, 492). Further on, he wondered whether 'some intelligent creatures, of different nature from the human, happen to inhabit a globe which is in that way animated, or will inhabit it' (Kepler 1937–, VI, 366, 1997, 495). The discussion then broadened to incorporate the nature of souls and minds elsewhere in the cosmos. He invoked Proclus' hymn and suggested that if one were to drink too deeply from the cup of Pythagoras, he might begin to dream that there were other reasoning faculties dispersed among the globes which orbit the sun. The faculty granted to the middle globe (the earth) is the most outstanding and absolute, while 'on the Sun dwells simple Understanding, the "intellectual fire" or "mind"' (Kepler 1937–, VI, 367, 1997, 496).[64] Finally, Kepler compared the variety observed on the surface of the sun, such as the recently detected sun-spots, to the meteorological activity of the earth, suggesting that it may serve a similar purpose in fostering life. 'What use is this furnishing, if the globe is empty?', he asked: 'Do not the very

---

[63] See, for example, Caspar's reflection on Kepler's harmonic philosophy: 'Certainly for Kepler this book was his mind's favorite child. Those were the thoughts to which he clung during the trials of his life and which distributed light to him in the darkness which surrounded him. They formed the place of refuge, where he felt secure, which he recognized as his true home.' Caspar 1993, 288. See also Field 2009.

[64] The editors of the English version of the *Harmony* remark that Kepler would have come across Proclus' hymn to the sun printed as an addition to the Orphic hymns. Kepler 1997, 493, n. 176. On Proclus' hymns, see van den Berg 2001, esp. 145–150 for the hymn to the sun.

senses themselves cry out that fiery bodies inhabit it, which have the capacity for simple minds?' (Kepler 1937–, VI, 368, 1997, 497).

Invoking again the opinion of Tycho Brahe as a precedent, Kepler argued that life should in fact be spread amongst all the globes of the world.[65] The ubiquity of life on earth, with species fitted to survive in various environments—be it aquatic, aerial, snowy or arid—suggested that God would adorn the other globes with living creatures suited to their conditions. Each globe in fact is 'commended' by different properties: the period and eccentricity of their orbits, their proximity to the sun, the brightness of their bodies, and most interestingly perhaps of all, 'the properties of the figures on which every region is supported' (Kepler 1937–, VI, 367, 1997, 497). Just as an impetus to sexual reproduction is provided by the harmonic progenies of the pentagon, so is the entire globe of the earth characterised by this form of generation. In Kepler's scheme of solids, the dodecahedron, which encompasses the sphere of the earth, is masculine, while the icosahedron, which supports the earth from the inside, is feminine. The act of sexual procreation is thus expressed 'in the divine proportion of that marriage and its inexpressibility' (Kepler 1937–, VI, 367, 1997, 497). One could therefore (although Kepler doesn't) speculate on the nature of the inhabitants of each globe based on the figure or figures which determine the sphere of its orbit. Perhaps sexual reproduction is unique to the earth, and on the other planets things are generated asexually or spontaneously?

Book 5 of the *Harmonice mundi*, more so than the *Dissertatio*, proves that astrology provided an impetus for Kepler's considerations of ET life. In his cosmology, the celestial bodies are both habitats for living creatures and sources of influence in themselves, as their emanations are received and compared at the other globes. Prior to making the discoveries in Book 5, the sun is only ever discussed in terms of its life-giving, not life-harbouring, properties. The musical scale being played out by the planets, wherein lay the entire logic behind the determination of their motions, necessitated the existence of at least some perceptive soul in the sun, and quite possibly some form of solar beings. It should be remembered that harmonic ratios formed by the extreme motions of the planets was in fact the hypothetical third cause of astrological influence described by Kepler in his *De fundamentis* in 1601. Yes, Kepler restricted his terrestrial astrology to considerations of the aspects, but the earth was no longer the sole recipient of celestial influence. In Kepler's cosmology extraterrestrial life went hand-in-hand with extraterrestrial astrology.

Astrology answered the *cui bono* question for a large number of astronomical phenomena. In the *Astronomia nova*, for example, he criticised the hypothesis of latitudinal libration advanced by the contemporary astronomer Magini, citing that it calculated no conjunctions of Mars and the sun which did not take place 'through the body'. If this were true, Kepler argued, it 'would render vain nature's latitudinal

[65] There is no textual evidence that Tycho in fact believed in life in/on the other globes, although that doesn't rule out the possibility that it may have been discussed privately between the two astronomers. Dick argues that Kepler misconstrued a *reductio ad absurdum* argument made elsewhere by Tycho, i.e. the immense size of a heliocentric universe would seem to necessitate celestial inhabitants to make it useful. Celestial inhabitation is absurd, ergo heliocentrism is incorrect. See Dick 1982, 73–74.

temperings, which prevent the excessive arousal of the sublunary powers'.[66] A decade later, in part of his *Epitome of Copernican Astronomy*, astrology provided a possible reason behind the tricky business of the motion of the apsides.[67] This was a phenomenon that entailed the slow migration through the zodiac of a planet's furthest departure from the sun, or aphelion, in its elliptical orbit. Not being able to provide a physical cause for this process, or even to accurately calculate it, Kepler resorted to a more metaphysical speculation:

> But in the meantime, I would not rigidly deny that this effect can be a part of the design, so that it is not a consequence of necessity, or a mere consequence: because we are still ignorant of the magnitude of it. Then there will be room to speak about the final cause: to the final cause belongs the mutual tempering of the forces of libration, of the deflection of the threads, and of the revolution, in some fixed proportion: in order that, because the librations were prepared in order to set up the harmonies of the movements, any given harmony should not be born always in some one configuration of two planets, but in the succession of the ages would pass through absolutely all the configurations, and in order that thus all the harmonies of movements—which Book V of the Harmonies is about—should be mingled with all the harmonic configurations—which is the matter of Book IV of the Harmonies.[68]

So Kepler's explanation of the motion of the apsides was that, from this same grand seat of the sun, the harmonic influence of the motions and the harmonic influence of the aspects should be constantly shuffled together.

Going further, we may still ask ourselves *what* good is served by this complex astrological system. Taking the example of astrology and the weather, one could ask why the meteorological operation of the earth, powered internally, should be sensitive to astrological aspects. This particular question is one that Kepler answered in Book 4 of *Harmonice*:

> However, although this operation of the Earth's soul is perpetual, yet there was a need of some amounts in excess to be evaporated, not continuously for the whole of some period of time, but confined to definite days, so that from the abundance of vapours emitted to the outside the seasonal rains might be supplied, though with some sunny weather interspersed, in order to revive and water the surface of the Earth, so that fruits and food for living creatures might spring up from them.[69]

---

[66] Kepler 1937–, III, 140–41: 'Quod si verum sit, frustra natura temperamentum latitudinum confinxerit, ne corporalibus conjunctionibus crebro contingentibus nimiae essent exagitationes sublunarium virtutum.' Kepler 1992, 232.

[67] Here again we can draw a parallel to Rheticus' *Narratio prima*. See above, Sect. 3.3.

[68] Kepler 1937–, VII, 342: 'At non interim rigide negaverim, hunc effectum potius in consilii parte fuisse, ut non sit, vel non sit mera necessitatis appendix: quia huius quantitatem adhuc ignoramus. Tunc locus erit dicendae causae finalis: huc tendere contemperationem inter se virium, librationis, fibrarum inclinationis, circumlationis, certa in unoquolibet proportione: ut quia librationes quidem comparatae sunt ad constituendas Harmonias motuum, Harmoniarum quaelibet enasceretur non semper in una aliqua binorum planetarum configuratione, sed successu saeculorum omnes omnino configurationes pervagaretur: atque sic Harmoniae motuum omnes (quae sunt lib. V. Harmonicorum) cum Harmoniis configurationum omnibus (libri IV. Harm: materia) permiscerentur.' Kepler 1952, 945.

[69] Kepler 1937–, VI, 272: 'Hoc vero Animae Terrae opus, etsi perpetuum est: opus tamen fuit excessibus aliquibus in evaporando; non continuis toto aliquo tempore: sed ad certos dies redactis; ut ex copia vaporum foras emissa pluviae tempestivae, Solibus tamen intercurrentibus, suppedita-

This is part of a fascinating issue in post-Aristotelian astrology, and indeed just a larger part of Kepler's own teleological project. Faced with the task of re-founding astrology within a non-geocentric framework, he not only had to answer 'how' celestial influence functioned, but 'why'. Here his answer was that the meteorological excitement accompanying the varied occurrence of astrological aspects shakes up what might otherwise be a monotonous weather cycle. The resulting daily variation of the excess and deficit of atmospheric vapours and rains is beneficial to the growth of vegetation on which animals depend. In conclusion, one could say that, for Kepler, celestial inhabitation stepped in as an explanation for phenomena when influence on the earth no longer seemed tenable. Another way of thinking about this, however, is to say that the posited inhabitation of the other celestial bodies allowed Kepler to explain *more* things according to a principle of influence—it's just that that influence was no longer solely directed towards the earth.

4.8 The *Somnium* and Kepler's Legacy to Pluralism

Kepler's *Somnium*, published posthumously in 1634, was an accumulative work, the basis of which was his lunar disputation from his student days at the University of Tübingen. The narrative framework was added later, and a version of this circulated in manuscript form, to which Kepler continued to add notes between 1620 and 1630. This was then supplemented with a geographical appendix, most likely the 'lunar geography' which, as noted above, he had written for his friend Wackher.[70] The purpose of the work as a whole was to present a lunar astronomy in a way which would remove any objections to the motion of the earth. The narrative itself is much shorter than the accompanying notes, and most of this is dedicated to a description of astronomical phenomena from a lunar perspective. What concerns us here is that smaller proportion of the work which is dedicated to considerations about the lunar inhabitants themselves, because it is probably the most detailed early example of the application of terrestrial biological concepts to the question of life on another celestial body. Kepler divided his lunar inhabitants into two main categories. The earth was known to the lunarians as *volva*, and those creatures which lived on the side of the moon always facing the earth he called 'Subvolvans'; those on the dark side of the moon he called 'Privolvans'. Both possessed what could be considered 'civilisation'.

Kepler believed that the circular cavities observable on the surface of the moon could not be produced by any natural motion of the elements, but must be the work of an intelligent mind. In the *Harmonice* he had quoted a passage from Proclus' *Commentary on Euclid* to the effect that nature does not produce perfectly straight

rentur, superficiei Terrae refocillandae humectandaeque causa; unde fruges et pabula Animantibus succrescere possent.' Kepler 1997, 368.

[70] Kepler, *Somnium, seu opus posthumum de astronomia lunari* (1634), in Kepler 1937–, XI.2, 317–438. Translations from Kepler 1967. On the composition of the work, see the introduction to Rosen's translation: Kepler 1967, xvii–xxiii.

lines, flat surfaces, right angles, or geometric shapes (Kepler 1937–, VI, 219; Proclus 1873, 14). In the geographical appendix he built on this argument, considering as axiomatic the proposition that order is always the product of mind. 'If you direct your mind to the towns on the moon', he entreated the reader, 'I shall prove to you that I see them.'[71] The suggestion is that the circular cavities on the moon have been created by the Selenites to provide shelter both from the elements and also from incursions by hostile forces. For once you institute a comparison between lunar and terrestrial populations, Kepler thought, you can make judgments about similar things. The circular settlements must be inhabited by different groups, while the rough surrounding regions must be host to 'wild and savage bands of thieves' (Kepler 1937–, XI.2, 376, 1967, 169).

This may seem like a somewhat naïve assessment of lunar craters, shaped by teleological and metaphysical biases, but even today the search for ET life and ET intelligence follows similar methodological principles. Astronomers look for phenomena that are out-of-the-ordinary, such as high concentrations of atmospheric oxygen or repetitive bursts of electromagnetic radiation, which may indicate a divergence from the normal or non-biological course of nature.[72] On the surface of the moon Kepler saw a natural elemental world, similar to the earth with many of its features, which had been shaped into order by the art and artifice of reasoning creatures. Even in this basic distinction between natural chaos and intelligent planning, he perceived the manifestation of universal providence. 'Indeed, the surface of the spheres seems to have been left to blind chance for this very purpose,' he proposed, 'that in arranging it and embellishing parts of it there may be room for the exercise of reason.'[73]

We have already seen how in earlier works Kepler imagined the lunar inhabitants to suit their environment, and at the same time imagined the lunar environment to suit its inhabitants. He took both approaches in the *Somnium*, asserting again his belief in the necessity of a lunar atmosphere and exhalations (Kepler 1937–, XI.2, 348, 1967, 89). He also maintained that the moon must be perforated with caves, especially on the Privolvan side, where the conditions were more extreme without the tempering influence of the earth, and the inhabitants in more dire need of shelter (Kepler 1937–, XI.2, 362, 1967, 128–29). Indeed, the environment on the Privolvan side of the moon required a liberal use of Kepler's imagination to guarantee inhabitation:

> Since I had deprived the Privolvans of water, and I was compelled to leave them immense alternations of heat and cold coming directly on each other's heels, it occurred to me that those regions could not be inhabited, at least not out in the open. Hence it was convenient that water flowed in at fixed times of the day. When it receded, I had the living creatures accompany it. To enable them to do so quickly, I gave some of them long legs and others the ability to swim and endure the water, with the proviso that they would not degenerate (*degenerare*) into fishes. None of this will be unbelievable to anybody who has read about

[71] Kepler 1937–, XI.2, 368: '[Ad] Oppida Lunaria si mentem afferas, tibi me videre comprobabo.' Kepler 1967, 151.

[72] Some methodological handbooks are Seager 2010; Shuch 2011.

[73] Kepler 1937–, XI.2, 373: 'Equidem ob id ipsum superficies globorum videtur permissa coeco casui, ut in ea. ordinanda et ornanda per partes, locus esset exercitio rationis.' Kepler 1967, 161.

> Cola, the Sicilian man-fish. Moreover, I thought that nothing on earth is so fierce that God did not instil resistance to it in a particular species of animals: in lions, to hunger and the African heat; in camels, to thirst and the vast deserts of Palmyrene Syria; in bears, to the cold of the far north, etc.[74]

The use of the term *degenerare* is interesting here. Fish were a category of life lower than that of other animals, being a product of spontaneous generation and therefore more comparable to insects. By referencing the legend of Cola, the Sicilian man who could survive underwater for hours at a time like a fish, Kepler made it clear that he wished his lunar inhabitants to be of a certain class or nobility of creature.[75]

Yet these lower forms of life exist in the moon as well. 'Things born in the ground', he imagined, 'generally begin and end their lives on the same day, with new generations springing up daily.'[76] The lunar day, of course, is the length of 15 earth-days, as is the night. The lunar flora could not possibly survive such a long time deprived of light and heat, and so the body of the moon, possibly carrying the seeds, brings forth a new generation following the eventual return of the sun. Furthermore, there are examples of bizarre generative processes:

> To certain of them the breath they exhaust and the life they lose on account of the heat of the day return at night; the pattern is the opposite of that governing flies among us. Scattered everywhere on the ground are objects having the shape of pine cones. Their shells are roasted during the day. In the evening when, so to say, they disclose their secrets, they beget living creatures.[77]

The note to this section expands, citing terrestrial examples:

> From the resin which exudes from ship timbers as a result of the sun's heat and sticks together in a ball, ducks are born. The last part of their whole body to develop is the bill. When this is released, they slip into the water below. So says Scaliger in his Exercises. The Scottish tree which produces the same offspring is well known because many people have made it famous. In the year 1615, when the summer was very dry, at Linz I saw a juniper twig that had been brought in from the abandoned fields of the Traun. The twig had given birth to an insect of unfamiliar shape and of the colour of a horned beetle. The insect had

[74] Kepler 1937–, XI.2, 363: 'Cum aquas Privolvis ademissem, aestus et frigoris vicissitudines immanes proximis omnino temporum articulis relinquere cogerer: subiit animum, eos tractus habitari non posse, saltem sub dio. Ex opportuno igitur aquae influebant certis diei tempestatibus: eas iterum recedentes jussi viventia comitari; et ut tam propere possent, longos pedes dedi; aliis nandi facultatem, et aquarum patientiam: ut tantum non in pisces degenerarent. Nec quicquam huius illi erit incredibile, qui de Cola Siculo, Homine-pisce, legerit. Etiam illud reputabam; nihil in terris tam esse nobis violentum, cuius tolerantiam non indiderit Deus certo generi animalium; famis et aestus Africani, Leonibus; situs et immensorum desertorum Syriae Palmyrenes, Camelis; frigoris Hyperborei, Ursis, etc.' Kepler 1967, 130–31.

[75] Kepler may have read about Cola in Scaliger 1557, Ex. 262. See Kepler 1967, 131, n. 353.

[76] Kepler 1937–, XI.2, 330: 'Terra nascentia … plerumque eodem die et creantur et enecantur, novis quotidie succrescentibus.' Kepler 1967, 28.

[77] Kepler 1937–, XI.2, 330: 'Quibusdam per diei aestum spiritus exhaustus, vitaque extincta, per noctem redeunt, contraria ratione quam apud nos Muscis. Passim per solum dispersae moles figura nucum pinearum, per diem adustis corticibus, vesperi quasi reclusis latebris, animantia edunt.' Kepler 1967, 28.

emerged up to its middle and moved slowly. The back parts clinging to the tree were juniper resin.[78]

Lunar creatures born from cone-shaped objects are examples of that natural process which brings forth living creatures from seemingly inanimate matter. Kepler gave earthly examples like ducks born from pitch or from trees, and stick insects (we may assume) from sticks. These are all manifestations of that universal faculty which is responsible also, as in *De stella nova*, for the production of comets, new stars, and the like. Lunar life, as depicted in the *Somnium*, demonstrates not only God's ingenuity when it comes to the coupling of life with its environment, but also the ability of nature to produce life itself in certain conditions. The moon is another example of Kepler's vision of a beautifully varied cosmos, but one which is united by certain universal laws and operations, and therefore to a degree understandable by recourse to terrestrial analogies.

The *Somnium* contains no discussion of astrology and therefore it is not surprising that, as the most famous example of his pluralist philosophy, historians have failed to make the connection between astrology and ET life in Kepler. It is another example of the danger of focusing on the most novel or interesting components of Kepler's philosophy without appreciating those components in the larger context of his work.[79] The *Somnium* is narrow in intention and therefore in scope, and for Kepler the question of life on the moon had always been tied together with concepts derived from his interpretation of Copernicanism, such as the relativity of motion and the unity of terrestrial and celestial nature. The expansion of life *beyond* the moon is linked closely in Kepler with the discovery of celestial influence in operation at non-terrestrial locations, especially the new star in the firmament and the harmonic motions of the planets around the sun. The particular and idiosyncratic nature of Kepler's astrology should not distract from his involvement in the larger trend outlined thus far. It is the same line of thinking visible in Plutarch, Cusa, Benedetti, Patrizi and Bruno (see Chap. 2): the role that the celestial bodies play in generative processes here on earth may be replicated between those bodies themselves, for similar purposes and with similar results. For Kepler, whose work was so influential on the development of the 'plurality of worlds' philosophy, the notion of life, both on earth and elsewhere, was deeply attached to the notion of influence from the celestial realm.

---

[78] Kepler 1937–, XI.2, 364: 'Ex resina, exsudante ex trabibus navium per Solis fervorem, et globatim adhaerescente, Anates nasci, quibus ultimum totius corporis, rostrum maturescat, quo soluto, se dent undis subjectis, refert Scaliger in Exercitationibus. Nota est multorum celebratione, arbor Scotiae, quae eundem foetum proferat. Anno 1615 aestate siccissima, vidi Lincii allatum ex Drani campis desertis, ramulum juniperi, cui adnata erat figura insecti insolita, colore scarabaei cornuti, mediotenus extans, et sese movens lento motu, posteriora arbori adhaerentia, erant resina juniperina.' Kepler 1967, 133–34.

[79] See Caspar's similar criticism of the historical treatment of Kepler's laws of planetary motion: 'In truth, if a work presents science with such a valuable contribution as the third planet law (not to mention the mathematical and musical fruits), then a critic must seek the lack in himself if he does not achieve an understanding of the manner of contemplating nature out of which the work has arisen.' Caspar 1993, 289.

It is tempting to see Kepler as a nexus between the histories of astrology and pluralism. He was the last great reformer of astrology who could still be considered part of the scientific elite, and interest in his astrological theory and practice continues to grow.[80] He is also one of the main characters in the history of pluralism in the seventeenth century. The *Somnium* is largely responsible for this, both because of the astronomical expertise it brought to the subject of lunar life, but also because the narrative setting places it as an early example or precursor of the literary genre of science fiction. Tales of voyages to the moon and the planets, many of them influenced by Kepler, would proliferate in the decades and centuries after Kepler, and they played a large part in disseminating and popularising theories of other worlds and ET life.[81] There has been very little mention, however, of the relationship between astrology and pluralism in Kepler.[82] Many would see it as another example of his Janus-like nature, looking both forward and back.[83] In reality, these were both ideas of contemporary interest with long histories, and they fit together quite harmoniously in Kepler's mind.

Appreciating this is crucial for the construction of an accurate and rich history of astrobiology—a history to which Kepler is already closely tied. This connection has been strengthened by the fact that his name was given to an orbital telescope which has been used to identify thousands of planets orbiting other stars (commonly called exoplanets). The irony in naming this telescope after a man who was vocally opposed to the idea that the stars could have their own planets has largely been lost, and besides is a trifling point.[84] What is more important for the history of astrobiology is to appreciate the pervasiveness of astrological ideas within Kepler's thought, and in particular how they contributed to the development of pluralist theories. Astrology was a significant factor in how the life on any particular celestial body was intimately related to that body's cosmic environment. In the following chapter, we will look at examples further into the seventeenth century of cosmologies which combined elements of astrology and pluralism, before turning our attention to an instance of these ideas under debate within the context of early mechanism.

[80] The scientific elite, that is, as it is construed by our modern history of science. There were later reformers of astrology, such as Jean-Baptiste Morin (1583–1656), who were considered elite in their own time.

[81] Over 10% of Dick's *Plurality of Worlds* (Dick 1982), which stretches from Democritus to Kant, is taken up by discussion of Kepler. The heavy-weighting given him in this standard text, as well as his prominent position in the history of astronomy more generally, means that Kepler is one of the first names mentioned in any short description of the history of pluralism, e.g. Catling 2013, 3–4.

[82] Dick did say this much: 'Finally, Kepler the astrologer, who had cast many horoscopes for Rudolph II, noted that the astrological point of view also favored the existence of Jovians, because the small arcs that the orbits of the new moons subtended could be of significance only to the Jovians.' See Dick 1982, 77. Kepler didn't explicitly link these moons to a Jovian astrology (see Boner 2013, 152), but it is not necessarily a misguided connection on Dick's part.

[83] See, for example, the analysis of the *Somnium* in Hallyn 1990, 279–80.

[84] He may well have been delighted, however, by the discovery that some moons and planets, in this system and others, have an orbital resonance, where their orbital periods are related by ratios of small integers. One can imagine the ambivalence he might have felt towards such articles as Holman et al. 2010, 'Kepler-9: a system of multiple planets transiting a sun-like star, confirmed by timing variations'.

References

Bialas, Volker. 2009. Kepler's philosophy of nature. In *Johannes Kepler: From Tübingen to Żagań*, ed. Richard Lynn Kremer and Jarosław Włodarczyk, 29–40. Warsaw: Institut Historii Nauki PAN.

Boner, Patrick J. 2007. Kepler v. the Epicureans: Causality, coincidence and the origins of the new star of 1604. *Journal for the History of Astronomy* 38: 207–221.

———. 2008. Life in the liquid fields: Kepler, Tycho and Gilbert on the nature of the heavens and Earth. *History of Science* 46: 275–297.

———. 2011. Kepler's Copernican campaign and the new star of 1604. In *Change and continuity in early modern cosmology*, ed. Patrick J. Boner, 93–114. Dordrecht: Springer.

———. 2013. *Kepler's cosmological synthesis: Astrology, mechanism and the soul*. Leiden: Brill.

———. 2014. Statesman and scholar: Herwart von Hohenburg as patron and author in the republic of letters. *History of Science* 52: 29–51.

Caspar, Max. 1993[1959]. *Kepler*. Ed. and Trans. Clarrise Doris Helmman. New York: Dover Publications.

Catling, David C. 2013. *Astrobiology: A very short introduction*. Oxford: Oxford University Press.

Crowe, Michael J. 1986. *The extraterrestrial life debate, 1750–1900: The idea of a plurality of worlds from Kant to Lowell*. Cambridge: Cambridge University Press.

Dick, Steven J. 1982. *Plurality of worlds: The origins of the extraterrestrial life debate from Democritus to Kant*. Cambridge/New York: Cambridge University Press.

———. 1996. Other worlds: The cultural significance of the extraterrestrial life debate. *Leonardo* 29: 133–137.

Dunér, David. 2016. Swedenborg and the plurality of worlds: Astrotheology in the eighteenth century. *Zygon* 51: 450–479.

Field, J.V. 1984. A Lutheran astrologer: Johannes Kepler. *Archive for History of Exact Sciences* 31: 189–272.

———. 1988. *Kepler's geometrical cosmology*. London: Athlone.

———. 2009. Kepler's harmony of the world. In *Johannes Kepler: From Tübingen to Żagań*, ed. Richard Lynn Kremer and Jarosław Włodarczyk, 11–28. Warsaw: Institut Historii Nauki PAN.

Gingerich, Owen. 1973. From Copernicus to Kepler: Heliocentrism as model and as reality. *Proceedings of the American Philosophical Society* 117: 513–522.

Grafton, Anthony. 1973. Michael Maestlin's account of Copernican planetary theory. *Proceedings of the American Philosophical Society* 117: 523–550.

Granada, Miguel A. 2005. Kepler v. Roeslin on the interpretation of Kepler's nova: (1) 1604–1606. *Journal for the History of Astronomy* 36: 299–319.

———. 2009. Kepler and Bruno on the infinity of the universe and of solar systems. In *Johannes Kepler: From Tübingen to Żagań*, ed. Richard Lynn Kremer and Jarosław Włodarczyk, 131–158. Warsaw: Institut Historii Nauki PAN.

———. 2011a. Johannes Kepler and David Fabricius: Their discussion on the nova of 1604. In *Change and continuity in early modern cosmology*, ed. Patrick J. Boner, 66–92. Dordrecht: Springer.

———. 2011b. After the nova of 1604: Roeslin and Kepler's discussion on the significance of the celestial novelties (1607–1613). *Journal for the History of Astronomy* 42: 353–390.

Grant, Edward. 1981. *Much ado about nothing: Theories of space and vacuum from the Middle Ages to the scientific revolution*. Cambridge/New York: Cambridge University Press.

———. 1994. *Planets, stars, and orbs: The medieval cosmos*, 1200–1687. Cambridge: Cambridge University Press.

Greenbaum, Dorian Gieseler. 2015. Kepler's personal astrology: Two letters to Michael Maestlin. In *From Māshā' Allah to Kepler: Theory and practice in medieval and Renaissance astrology*, ed. Charles Burnett and Dorian Gieseler Greenbaum, 177–200. Ceredigion: Sophia Centre Press.

Hallyn, Fernand. 1990. The poetic structure of the world: Copernicus and Kepler. In *New York*. Cambridge, MA: Zone Books; Distributed by MIT Press.

Hirai, Hiro, ed. 2008. *Cornelius Gemma: Cosmology, medicine, and natural philosophy in Renaissance Louvain*. Pisa: Fabrizio Serra.
———. 2011. *Medical humanism and natural philosophy: Renaissance debates on matter, life and the soul*. Leiden: Brill.
Holman, Matthew J., Daniel C. Fabrycky, Darin Ragozzine, Eric B. Ford, Jason H. Steffen, William F. Welsh, Jack J. Lissauer, et al. 2010. Kepler-9: A system of multiple planets transiting a sun-like star, confirmed by timing variations. *Science* 330: 51–54.
Jardine, Nick. 2009. God's "ideal reader": Kepler and his serious jokes. In *Johannes Kepler: From Tübingen to Żagań*, ed. Richard Lynn Kremer and Jarosław Włodarczyk, 41–52. Warsaw: Institut Historii Nauki PAN.
Juste, David. 2010. Musical theory and astrological foundations in Kepler: The making of the new aspects. In *Music and esotericism*, ed. Laurence Wuidar, 177–196. Leiden: Brill.
Kepler, Johannes. 1937. *Gesammelte werke. Edited under the supervision of the Bayerische Akademie der Wissenschaften, 25 vols published so far*. Munich: C.H. Beck.
———. 1952. Epitome of Copernican astronomy: IV and V. Translated by Charles Glenn Wallis. In *Great books of the Western world, volume 16: Ptolemy, Copernicus, Kepler*, ed. Robert Maynard Hutchins. Chicago: Encyclopædia Britannica; W. Benton.
———. 1965. *Kepler's Conversation with Galileo's Sidereal Messenger*. Trans. Edward Rosen. New York: Johnson Reprint Corp.
———. 1967. *Somnium: The dream, or Posthumous Work on Lunar Astronomy*. Trans. Edward Rosen. Madison: University of Wisconsin Press.
———. 1981. *Mysterium Cosmographicum: The Secret of the Universe*. Trans. A.M. Duncan. New York: Abaris.
———. 1992. *New Astronomy*. Trans. William H. Donahue/Cambridge/New York: Cambridge University Press.
———. 1997. *The Harmony of the World*. Trans. E.J. Aiton, A.M. Duncan, and J.V. Field. Philadelphia: American Philosophical Society.
Kremer, Richard Lynn, and Jarosław Włodarczyk, eds. 2009. *Johannes Kepler: From Tübingen to Żagań*. Warsaw: Institut Historii Nauki PAN.
Lindberg, David C. 1986. The genesis of Kepler's theory of light: Light metaphysics from Plotinus to Kepler. *Osiris* 2: 4–42.
Losch, Andreas. 2016. Astrotheology: On exoplanets, Christian concerns, and human hopes. *Zygon* 51: 405–413.
Martens, Rhonda. 2000. *Kepler's philosophy and the new astronomy*. Princeton/Oxford: Princeton University Press.
Methuen, Charlotte. 1996. Maestlin's teaching of Copernicus: The evidence of his university textbook and disputations. *Isis* 87: 230–247.
———. 1998. *Kepler's Tübingen: Stimulus to a theological mathematics*. Aldershot/Brookfield: Ashgate.
Mix, Lucas John. 2016. Life-value narratives and the impact of astrobiology on Christian ethics. *Zygon* 51: 520–535.
Paul, Robert. 1986. Joseph Smith and the plurality of worlds idea. *Dialogue: A Journal of Mormon Thought* 19: 13–36.
Proclus. 1873. *In primum Euclidis Elementorum librum commentarii*, ed. Gottfried Friedlein. Leipzig: Teubner.
Rabin, Sheila J. 1997. Kepler's attitude toward Pico and the anti-astrology polemic. *Renaissance Quarterly* 50: 750–770.
Rutkin, H. Darrel. 2005. Galileo astrologer: Astrology and mathematical practice in the late-sixteenth and early-seventeenth centuries. *Galilaeana* 2: 107–143.
———. 2006. Various uses of horoscopes: Astrological practices in early modern Europe. In *Horoscopes and public spheres: Essays on the history of astrology*, ed. Günther Oestmann, H. Darrel Rutkin, and Kocku von Stuckrad, 167–182. Berlin/New York: Walter de Gruyter.
Scaliger, Julius Caesar. 1557. *Exotericarum exercitationum liber quintus decimus de subtilitate ad Hieronymum Cardanum*. Paris: Michel de Vascosan.

Seager, Sara. 2010. *Exoplanet atmospheres: Physical processes*. Princeton: Princeton University Press.

Shuch, H. Paul, ed. 2011. *Searching for extraterrestrial intelligence: SETI past, present, and future*. Chichester: Springer.

Simon, Gérard. 1975. Kepler's astrology: The direction of a reform. *Vistas in Astronomy* 18: 439–448.

Snyder, James G. 2011. Marsilio Ficino's critique of the Lucretian alternative. *Journal of the History of Ideas* 72: 165–181.

Stephenson, Bruce. 1987. *Kepler's physical astronomy*. New York: Springer.

———. 1994. *The music of the heavens: Kepler's harmonic astronomy*. Princeton: Princeton University Press.

van den Berg, Robbert Maarten. 2001. Proclus' hymns: Essays, translations, commentary. In *Leiden*. Boston: Brill.

Voelkel, James R. 2001. *The composition of Kepler's Astronomia nova*. Princeton: Princeton University Press.

Walker, D.P. 1967. Kepler's celestial music. *Journal of the Warburg and Courtauld Institutes* 30: 228–250.

Chapter 5
Influence and/or Inhabitation: The Celestial Bodies Between Kepler and Newton

> *Ramus did overthrow Aristotle's Philosophy; Copernicus, Ptolomey's Astrologie; Paracelsus, Galen's Physick; So that every one hath followers and disciples, and all appearing plausible. We have much ado whom to believe, and thereby are constrained to confesse, that what we know, is much lesse then what we know not.*
>
> (Borel 1658, 6)

Abstract After the case studies of the previous chapters, this chapter presents a selected survey of seventeenth-century philosophers with an eye on the developing relationship between astrological and pluralist theories. It focuses first on the continuing persistence of astrological thinking within pluralist frameworks, arguing that a symbiosis of sorts arose between theories of celestial influence and celestial inhabitation. It then turns to the early mechanical philosophies of Kenelm Digby, Thomas White and Thomas Hobbes. While looking at the relevance of mechanical principles to questions of celestial influence and inhabitation, this section also turns to second argument of the book: the growing antagonism between astrology and pluralism in terms of celestial teleology.

Keywords Astrology · Extraterrestrial life · Copernican · Cartesian · Nicholas Hill · Philippe van Lansberge · Henry More · Otto von Guericke · Claude Gadroys · Mechanical philosophy · Kenelm Digby · Thomas White · Thomas Hobbes

© Springer Nature Switzerland AG 2019

J. E. Christie, *From Influence to Inhabitation*, International Archives of the History of Ideas Archives internationales d'histoire des idées 228,
https://doi.org/10.1007/978-3-030-22169-0_5

5.1 Introduction

The preceding chapters have attempted to show how a belief in ET life grew within and was in fact spurred on by the dominant astrological paradigm. The philosophies of Gilbert and Kepler, and those who preceded them, hinted at a universe with multiple inhabited globes, all participating in a mutual exchange of virtues and influences. We will come across similar cosmological themes in the following chapter, but our attention will also shift onto how these ideas could be placed in opposition. This will then serve as groundwork for the second main argument of this book: that the theory of celestial inhabitation came to be placed purposefully in opposition to theories of celestial influence, a process which led to the establishment of pluralism as the central paradigm of astronomical cosmology. The main issue is of course one of teleology, and the next two chapters will sketch out the process by which these two ideas went from being teleologically complementary to teleologically opposed. This development was heavily tied to the obsession with atheism amongst philosophers in this period, natural or otherwise. Whereas in earlier periods astrology was often used as a defence against the perceived atheism of atomism, in the seventeenth century we see the situation almost reversed, with forms of pluralism and atomic vitalism being used to combat the threat of an astrological cosmology portrayed as deterministic and atheistic.

Apart from Giordano Bruno, about whose philosophy we said relatively little, none of the figures studied so far have been atomists. It was, however, the resurgence of atomistic or corpuscularian ideas which formed the basis for the mechanical philosophies of the mid-seventeenth century, and it was these mechanical philosophies, Cartesianism in particular, which nurtured and propagated pluralism. Implicit in early atomism and mechanism, treating physical phenomena as the motion and collision of effectively indivisible particles, was the rejection of the terrestrial/celestial division. Most mechanical philosophers were Copernicans, and so believed the earth to be one of the celestial globes, which, as we have already seen, could have important impacts on notions concerning the activity and/or passivity of those bodies. At the same time, there was a desire within mechanical philosophy to do away with occult qualities, and astrology saw itself largely excluded or marginalised. It is not the aim here to delve into the myriad of factors contributing to the decline of astrology as a scientific discipline, although certain aspects, such as the appropriation and rhetorical rebranding of astrological concepts by the new philosophies, will form part of the discussion.[1] The intention is rather to suggest that pluralism has an important and hitherto under-appreciated role in this narrative, and to that end this chapter will compare and contrast the views of individual philosophers concerning the compatibility or incompatibility of astrology and pluralism within a more mechanical cosmos.

---

[1] On this see Hutchison 1982; Millen 1985; Henry 2008. On the mechanisation of celestial influence, see also Roos 2001, 165–220.

To begin, however, it is worth touching briefly on a few developments in the first four decades of the seventeenth century that were passed over while our attention was on Gilbert and Kepler. One of the earliest post-Brunian attempts to describe an atomistic, pluralistic cosmos was made by the English philosopher Nicholas Hill (1570–c. 1610) in his *Philosophia Epicurea* (1601).[2] Hill was a member of the Northumberland Circle, and his work was influenced not only by Bruno but also by Gilbert, as well as Democritus, Hermes Trismegistus and Aristotle.[3] The two championed novelties of his work are atoms and the infinity of the world, but magnetism, and effused substantial forms and species more generally, take on a large role in the functioning of this new cosmos (Clucas 1995, 38). In Hill's infinite universe the reality of superiority and inferiority is abolished, and they are therefore considered as respective things. He proposed magnetic attraction as an analogue for the return of projected earthly bodies (N. 205, Hill 1619, 55–56). Magnetic laws also explained why the primary globes 'fall' (*decumbere*) towards one place more than other, while the stable distances between them prevents any possible coming together (N. 438, Hill 1619, 160). The similarity between the globes extends beyond magnetism to include their material and indeed inhabitants:

> The superior globes (as many call them) are of the same material as the globe in which we live, and in which all things are analogous to ours, the globes being super-fertile, and distinct species being subordinated to and ruling one another. Perhaps there are giant men in the sun of a size exceeding ours 160 times, compared with which we are dwarfs and slave-boys, while the lunar men are perhaps pygmies even smaller again.[4]

Hill anticipated the concern that this infinite world, in which the observable universe is an insignificant part, may be detrimental to man's perceived standing. In his view, the reason for this concern is that we are more likely to be envious of the humanity of another than to appreciate our own blessings. 'The special prerogative of our nature', he reassured us, 'in no way suffers from the propagation of numerous men.'[5]

He even suggested that God separated the orbs from one another in order to guard against human envy, but this separation (*disterminare*) does not prevent some form of interaction between the globes.[6] There are many astrological arguments in

---

[2] References are to the second edition, Hill 1619. The work consists of approximately 522 inconsistently numbered propositions, signified hereafter by 'N'.

[3] On Hill's cosmology, see Jacquot 1974. See also McColley 1939; Kargon 1966, 5–17, 43–53; Massa 1977; Dick 1982, 48–50; Trevor-Roper 1989; Clucas 1997, 266–68.

[4] N. 278, Hill 1619, 79–80: 'Globi superiores (oportet loqui cum vulgo) eiusdem sunt materiae cum globa quam incolimus, in quibus etiam ea. sunt omnia quae apud nos secundum analogiam, globis superfoetantibus, speciebus distinctis sibiinvicem subordinatis et imperantibus tum illis, tum sibiinvicem hominibus gyganteis, mole exuperantibus nostrates centies sexagesies forsan qui in sole sunt, ad quos comparati homunciones nos sumus, et servoli, lunaribus hominibus pygmeis multo forsan humilioribus.'

[5] N. 482, Hill 1619, 187: '...addo naturae nostrae specialem praerogativam nullatenus sublatam numerosa hominum propagatione...'

[6] N. 482, Hill 1619, 187: '...qui (Deus) ut invidiam forsitan praecaveret orbes ita a seinvicem disterminavit'.

the *Philosophia Epicurea*, although in terms of judicial astrology he limited its foundations to the aspects and the 'manifest analogy of superior to inferior things'.[7] Elsewhere he described the powers of things (*vires rerum*) as 'the influxes and effusions of the combined specific forms of the planets, aspects and ascendants'.[8] We can see the stress on forms, powers, species and active virtues as compensation for the underdeveloped explanatory power of the nascent atomism, and as an early sign of the problems the mechanical philosophy would have with the declared passivity of matter.[9] Hill saw elements in atomic terms, but argued that they receive their strength 'from the firmament', and although the seeds of offspring gestate in the womb, they do not do so 'without celestial influences'.[10] Not only the growth of the individual creature but also the wellbeing of the earth itself as a habitat depends on celestial interaction:

> Orbs do not exhaust themselves, or empty themselves, but by mutual virtues revive themselves, and, once they join forces together, they grow strong by turns from the reciprocal communication of their peculiar properties (*idiomatum communicatio*), and refresh each other's weakened powers.[11]

This vision of harmonious discourse between the primary globes is of course very similar to that of Bruno and especially Gilbert. In Hill's cosmology the celestial/terrestrial divide is abolished, the elements are thought of in terms of atoms, the celestial medium is replaced by a vacuum, and yet the paradigm of celestial influence holds fast.

The key analogy was magnetism, and it was grounded again in emanation theory. More than just that, Hill saw the consideration of such active virtues as a way to understand more about the nature of the universe as a whole:

> The natural and usual action of the form of anything is a certain emanation. Operation is an indicator of virtue; virtue an indicator of substance. We have no more certain way of grasping the agreement of substance of innumerable worlds, or of discovering their virtue, than the perceived similarity of action.[12]

---

[7] N. 392, Hill 1619, 131: 'Illa sola astrologica iudicia aestimanda quae in globorum aspectibus primariorum fundatur, et superiorum ad inferiora manifesta analogia.'

[8] N. 252, Hill 1619, 67: 'Vires rerum sunt formarum specificarum, planetarum, aspectuum, ascendentiumque influxus, et effluxus coeuntium.'

[9] On the influence of al-Kindi, Grosseteste and Bacon on the *vis radiativa* theory favoured by members of the Northumberland Circle, see Clucas 2001, 182–201.

[10] N. 241, Hill 1619, 64: 'Elementa suas a firmamento vires nanciscuntur, et licet semen propaginis gestent in utero, tamen non generant absque firmamentalibus influxibus.'

[11] N. 249, Hill 1619, 67: 'Orbes non se exhauriunt, aut exinaniunt, sed mutuis viribus se refocillant, et confoederati alterna idiomatum communicatione seinvicem corroborant, et redintegrant fractas vires alter alterius.'

[12] N. 248, Hill 1619, 66–7: 'Naturalis, et Usitata cuiusque actio formae quaedam est emanatio, operatione virtutis, virtute substantiae indice, nec certius habemus ad mundorum innumerabilium consensum substantialem tenendum, aut inveniendum medium, quum actionum analogiam conspicuam.' Reading 'quum' as 'quam', and reading 'medium' as a reference to 'virtus' in the operation/virtue/substance hierarchy. 'Medium' could also be read as a noun referring to the intervening space between the worlds.

Assuming the similarity of material and physical law in the heavens and the earth, while retaining a belief in the influence of the celestial over the terrestrial, Hill saw a way to argue from effect to cause so as to discover the nature of the substance of the other globes themselves. This is an argument we have seen before, for example, when Plutarch argued that the moon must be wet because of its humidifying influence (see Sect. 2.1). The argument that the ontological superiority of the celestial bodies is demonstrated by their influence was flipped around by Hill, who, with magnetism as his prime example, argued that a similarity of effects entailed an agreement of substance.

There were other signs that the similarity of celestial to terrestrial material would not immediately undermine the theoretical foundations of astrology. The Italian philosopher Giulio Cesare Vanini (1585–1619), whose naturalistic and highly astrological interpretation of God's providence saw him accused of atheism, dismissed the Aristotelian notion that the two realms were made of different types of matter (Davidson 2005, 72). Vanini came to be greatly influenced by the philosophy of Pietro Pomponazzi (1462–1525) following his move from Naples to Padua. He was a materialist who strongly advocated a form of astral determinism, believing that the heavenly bodies were the tools with which God exercised his providence in terrestrial affairs (Maclean 2005). This was in itself not a particularly heterodox opinion—Melanchthon, as we have seen, argued very similarly (see Sect. 2.2)—but Vanini's contemporaries were appalled by the way he granted all agency to physical and natural causes, leaving nothing to the spiritual or supernatural. He accordingly denied that the celestial bodies possessed intelligences by which they moved. Rather, he argued that they were possessed of an appropriate form which produces circular movement, and he used the analogy of a clock to illustrate the possibility of a certain and enduring law of motion (Vanini 1616, 21). Better known for its adoption by the mechanical philosophy, it is worth remembering that the world-as-clock metaphor was used much earlier (and perhaps more accurately) by Johannes de Sacrobosco to describe the *machina mundi*, in which motion and action were transferred inwards to the earth from the law-defined motion of the heavens.[13]

This world picture was upturned by Copernicus, but, as we have seen, the paradigm of celestial influence could be salvaged from its original framework. Tommaso Campanella (1568–1639) was one philosopher whose enthusiasm for the astronomical discoveries of Galileo failed to conflict with his enthusiasm for astrology. Indeed, when he wrote to Galileo giving his thoughts on the *Sidereus nuncius*, he was particularly intrigued by the possibility of ET life and suggested that 'it should especially be investigated … to what extent the inhabitants of each star have knowledge of astronomy and astrology'.[14] On top of this, we have seen from Rheticus

[13] John of Sacrobosco, *On the Sphere*, quoted in Grant 1974, 465.

[14] Letter from Tommaso Campanella to Galileo (January 13th, 1611), in Galilei 1890–1909, XII, 22: 'Illud et maxime investigandum erat … qualem habent astrologiam et astronomiam singulorum incolae astrorum.' On Campanella's advocacy of pluralism, see Dick 1982, 90–93; Ernst 2015.

onwards that new astronomical concepts could be argued astrologically.[15] Such was the case with one of the first outspoken advocates of heliocentrism in the Netherlands, Philippe van Lansberge (1561–1632). His *Bedenckinghen op den daghelijckschen ende jaerlijckschen loop vanden aerdt-cloot* ('Considerations about the Daily and Yearly Movements of the Earth', 1629), translated into Latin the following year, contains many arguments from influence for the reality of the Copernican system.[16] Like Kepler, Lansberge used Trinitarian theology as a cosmological model, dividing the world into three heavens: the first planetary heaven, the second heaven of the fixed stars, and the third invisible heaven beyond that. Although distinct, he argued that the energies of these three heavens all worked together for the same goal (Lansberge 1630, 32–33). In his attempt to explain the advantages of the heliocentric arrangement, he appealed to both influence and inhabitation to make his case.

In terms of the positions of the sun and the earth, justifying the switch is quite easy, because the two most important bodies were simply swapping positions. Lansberge began with the sun:

> Firstly, however, I will deal with the sun, which God established in the centre of the First Heaven; and he did not do so without good cause. It is not 166 times bigger than the earth, as Ptolemy thought, but 433 or more times bigger, as I will show clearly in our *Uranometria*. And what better place could there be to establish this torch of such great size and inexhaustible light than in the centre, from whence it may illuminate the whole of the First Heaven on every side? Certainly, God established the sun in this place like a king on a throne, from whence he can behold and rule his subjects on all sides. *There is nothing hid from the sight and heat thereof* (Psalm 19:7), so that whatever lives under its rule is made a participant in its life-giving faculty, this according to the sentence of the Prince of Philosophers, where he asserts that the sun and man generate man. Indeed, unless the life-giving power of the sun contributed to the generation of man and the other animals, it would be useless to attempt any such generation.[17]

The argument is not difficult to make, and had applied equally as well when the sun was considered to be in the middle position among the planets. The implied consequence, however, is that the life-giving faculty (*vivifica facultas*), indispensable to the process of generation on earth, is dispersed to every body in the first heaven. Lansberge then focused on the privileged position of the earth. 'It [the earth] was

---

[15] Indeed, it is worth remembering that Thomas Digges (c. 1546–1595) presented the Copernican hypothesis to an English audience as an addendum to a new edition of his father's astrological almanac. See Sect. 2.4.

[16] Refences are to the Latin translation, Lansberge 1630. On Lansberge, see Vermij 2002, 73–97. On his theories of celestial influence in particular, see Vermij 2016, 300–302.

[17] Lansberge 1630, 38: 'Primo autem de Sole agam, quem Deus constituit in centro Primi Caeli; neque id sine gravi causa. Est enim Ille, non 166 vicibus, ut voluit Ptolemaeus; sed 433 et amplius, Terra major; ut in Uranometria nostra evidenter ostendam. Sed hanc tante magnitudinis, tamque inexhausti luminis, Facem; quis aptiori loco collocarit, quam in Centro; unde totum Primum caelum circumquaque illuminaret? certe ut Regem in Solio, Deus Solem isto loco collocavit: unde subjecta omnia, undique respiciat, & regat: a cuius intuitu & Calore, nihil queat abscondi, Psal. 19.7. ut quicquid sub ejus versatur imperio, Vivifica Facultatis fiat particeps, ex sententia Principis Philosophorum, qua Solem & Hominem, Hominem generare, asserit. Nisi enim Vivifica Solis Vis, ad generationem Hominis, aliorumque animalium concurreret; irritum esset, quicquid ad generationem molirentur.'

set, therefore, in the middle of the planets', he wrote, 'the most dignified place in the whole of the First Heaven, to which both the superior and inferior planets freely transmit their influence, as if to a common storehouse.'[18] Lansberge later expanded on this point, using the analogy of a balance to demonstrate that the middle, rather than the centre, is the most natural position for the earth to receive the beneficial influences of the planets:

> But it should be noted first of all, that the balance of God, by which he weighs the powers of the planets every day, was attached to the earth, not to the sun or the other planets. For the sphere of the earth is the middle point between the inferior and superior spheres, so that these are suspended like scales on either side. For which reason we teach that the proportion of the powers and operations of the planets, which God distributes every day like weights, is aimed at the convenience of the globe of the earth, and especially of man, the inhabitant of it. For indeed we have not yet detected any mutual alteration in the planets from each other; while this [alteration], however that may be, always descends to the earth, as will be apparent from the following.[19]

Lansberge then gave five examples of the tangible effects of conjunctions between superior and inferior planets, mainly in terms of temperature and weather. He was maintaining an orthodox opinion on the influential role of the planets in the new cosmology, even going as far as to describe how the balance is maintained, with each superior planet having its opposite, in terms of qualities and influences, among the inferior ones (Lansberge 1630, 48–49). Yet he didn't completely shut the door on the possibility that alteration might occur among the planets themselves, and indeed he had allowed it to the moon a few pages earlier when he had put forward three arguments for why the moon was similar to the earth. Firstly, 'as the moon is like another sun to the earth,' he argued, 'so is it like another earth to the sun'.[20] This is shown by its solidity and its monthly rotation, which exposes every part of it to the light of the sun. Secondly, its dark and light patches seem to show that it is constructed in a similar fashion to the earth. Thirdly, it seems to act upon the earth 'as if it were an earth itself', as evidenced by the tides.[21] The conclusion is that the moon is like another earth (*altera Terra*), and that these two celestial bodies 'operate mutually between themselves'.[22]

---

[18] Lansberge 1630, 41: 'Posuit ergo eam in medio Planetarum; loco totius Primi Caeli dignissimo: ad quem tam superiores, quam inferiores Planetæ, Influentias suas, quasi ad commune penu[s], liberaliter transmittunt.'

[19] Lansberge 1630, 49: 'Sed imprimis notandum, Deum O. M. Libram, qua Vires Planetarum quotidie expendit; appendisse Terrae; non Soli, aut alij Planetae. Nam Sphaera Terrae, media est inter Sphaeras inferiores, et superiores: ita ut inferiores Planetae, una superiores altera Lance, quasi suspendantur. Quo docemur, Temperamentum Virium et Operationum Planetarum, quas Deus indies tanquam ad pondus distribuit; tendere in commodum Globi Terrae, atque imprimis Hominis, eum incolentis. Nullam etenim in Planetis a se mutuo alterationem, hactenus deprehendimus: cum ea., quocunque tandem modo fiat, semper in Terram cadat; ut ex sequentibus erit manifestum.'

[20] Lansberge 1630, 46: 'Porro sicut Luna, Terrae velut alter Sol; ita Soli est, velut altera Terra.'

[21] Lansberge 1630, 46: 'Luna non aliter operatur in Terram, quam si ipsa Terra esset.'

[22] Lansberge 1630, 47: 'Patet ergo, Lunam quasi alteram Terram esse; et has in se mutuo operari.'

One more thing Lansberge had to explain in the heliocentric system was the greatly expanded distance between Saturn and the fixed stars, which, since Galileo, had themselves been shown to be much larger and more numerous. Here again, influence and inhabitation combined, although the Calvinist Minister was far from maintaining other worlds with human-like creatures. First, in a quite convoluted manner, he dealt with the light of the stars:

> The Second [Heaven] is illuminated by the fixed stars, which, no less than the sun, received particular and distinct light from God. But [that light] is as much stronger than that of the First Heaven as it [the Second Heaven] is bigger, for God made the light of each heaven proportional to its size. Hence we assert the light of the fixed stars to be circumscribed by its own limits; nor to be extended much further than to the sphere of Saturn. Indeed, around here [Saturn] the lights of the dual heavens come together, so that, united between themselves, by their conjoined powers, as much from below as from above, they extend their operation in the service of either heaven; but especially of man, who, as if Lord of the Universe, God set in the First Heaven.[23]

But on top of this attempt to rescue some use of the Second Heaven for humanity, Lansberge argued that it is not devoid of God's creatures:

> Although the space between the orb of Saturn and the fixed stars is huge, it is not a vacuum, as Tycho Brahe and his followers think. Rather, it is filled with God's creatures, as much as the First Heaven. For besides new stars arising there, it is everywhere occupied by a multitude of invisible creatures; such that neither Tycho Brahe nor anyone else has any occasion for suspecting it to be empty or vacuum.[24]

The Tychonic system, by keeping the earth stationary, had no need to assume such a vast space between Saturn and the fixed stars to account for the lack of observable stellar parallax. The enormity of this space was one of Tycho's most consistent criticisms of a heliocentric system, and it is significant that part of Lansberge's answer relied upon the idea of living-space. Of course, this is not pluralism in the form that has interested us thus far. These are angelic, incorporeal creatures, not mortal animals or humans, and they are living in the ether rather than calling another planet their home.

These examples demonstrate that, beyond Gilbert and Kepler, thinkers in the first half of the seventeenth century—even some with pluralist leanings—not only saw the benefit of astrological reasoning for justifying new cosmological theories, but were also perfectly capable of accommodating astrological ideas to those new sys-

[23] Lansberge 1630, 33: 'Secundum porro; illustratur a Fixis: quae non minus quam Sol, proprium ac distinctum a Deo Lumen accepere: sed tanto fortius, quanto majus hoc Coelum Primo, Deus enim O. M. Lumen cujusque Caeli, proportionale fecit eius Amplitudini: unde etiam Lumen Fixarum, suis terminis circumscribi asserimus; nec ulterius extendi, quam ad Sphaeram Saturni. Circa hanc enim concurrunt binorum Caelorum Lumina; ut inter se unita, conjunctis viribus, tam sursum quam deorsum, operationem suam extendant: in usum utriusque Caeli; sed imprimis Hominis, quem Deus tanquam Dominum Universi, in Primo Coelo collocavit.'

[24] Lansberge 1630, 53: 'Licet ingens sit spatium, intra Orbem Saturni, et Fixarum; non tamen est vacuum, ut Tycho Brahe cum asseclis, sensit; sed Creaturis Dei repletum, aeque ac Primum Caelum. Nam praeter Novas Stellas inibi enascentes; tanta multitudine Creaturarum invisibilium undique obsidetur; ut nec Tycho Brahe, nec quisquam alius, ansam habeat suspicandi, inane esse aut vacuum.'

tems. The 1630s, however, saw the publication of several important and influential works on pluralist themes that had little or nothing to do with astrology. One was *The Discovery of a World in the Moone* (1638) by John Wilkins (1614–1672), which relied mainly upon the Copernican view of the earth as a planet as well as the more particular arguments about the earth-like nature of the moon found in Plutarch, Kepler and Galileo.[25] Celestial influence plays very little part in its argumentation, although he did remark on the 'compendium' of providence that could make the same body a world for habitation and a moon for the use of others (Wilkins 1638, 43). Another work was the fictional tale of a journey to the moon written by Francis Godwin (1562–1633), published posthumously in the same year as Wilkins' book.[26] This work, along with Kepler's *Somnium* and the later *L'Autre Monde: Ou les États et Empires de la Lune* ('The Other World, or A Comical History of the States and Empires of the Moon', 1658) of Cyrano de Bergerac (1619–1655), is considered an example of early modern science fiction.[27] Combined with Wilkins' scientific treatise, all these works popularised the idea of an inhabited lunar world. Instead of the moon being a special case (as we saw in Patrizi, for example), these works demonstrate how the moon, which had always been an exemplar of celestial influence, was now being held up as an exemplar instead of broader celestial inhabitation. They will not, however, form a significant part of this discussion, partly because they have been the most commonly studied of all the various aspects of the history of pluralism, and also because they have no obvious or direct relationship to celestial influence or astrology.[28]

This chapter will instead continue to follow the interaction between pluralism and celestial influence further into the period from the 1640s to the 1680s, investigating examples of natural philosophers who built both ideas into their new cosmologies. This chapter and the next will, in addition, present instances where the merits of these ideas are debated and indeed pitted against each other. Towards the end of the century, the advocates and popularisers of the new science will largely restrict talk of influences to those of the sun(s) and moon(s), and pluralism will take on the majority of the teleological burden. The marginalisation of astrology, however, was not an immediate or necessary consequence of the decline of Aristotelianism and the rise of the mechanical philosophies. As the following discussion will demonstrate, there was not one 'astrology' or one 'pluralism' to be accepted or rejected; they were both cosmological putty in the hands of the community of seventeenth-century world-builders.

---

[25] See Dick 1982, 97–104.

[26] Godwin 1638. There are interesting cosmological features of Godwin's work that show the influence of Gilbert. See Roos 2001, 133–34; Hutton 2005; Poole 2005.

[27] For an introduction to this topic, see the chapter on 'Seventeenth-Century Science Fiction', in Roberts 2006, 36–63.

[28] See Nicolson 1948; Bachrach 1987; Romanowski 1998; Campbell 1999, 135–76; Roos 2001; Hutton 2005; Cressy 2006.

5.2 The Symbiosis of Influence and Inhabitation in the New Cosmologies

In this section, we will examine a selection of the new cosmologies which arose in the mid-seventeenth century with an eye on two factors in particular—the continuing persistence of astrological thinking within pluralist frameworks, and the way in which pluralism encroached on previously astrological domains. By the 1640s, Copernicanism had been widely disseminated and, more importantly, developed into a feasible world-system by the celestial physics of Kepler's *Epitome of Copernican Astronomy* (1617–1621) and Galileo's *Dialogue Concerning the Two Chief World Systems* (1632). This period also saw the development of the mechanical philosophies, epitomised by the Christian-Epicurean atomism of Pierre Gassendi (1592–1655) and the physics of René Descartes (1596–1650). The works of these figures stimulated much of the interest in ET life in the middle and later decades of the seventeenth century, but none of the thinkers we will examine adhered strictly to their principles.

The philosophy of the Cambridge Platonist Henry More (1614–1687) is central to the history of the plurality of worlds tradition. His poetic work *Democritus Platonissans* (1646), a syncretic mix of Cartesianism, Platonism and Christian theology, has been awarded a prominent role in the development of pluralism and sufficiently well analysed in this respect that we need only mention a few of its relevant facets before moving on.[29] Henry More argued for that now familiar Copernican consequence that whatever is done on earth is done similarly in the other globes (More 1646, 4). He also argued that the stars in the sky have little relevance to us in terms of influence (More 1646, 6). Rather, the light any star shines to us in the evening is secondary to its main duty, which is to supply vital heat to its surrounding planets and to raise 'long hidden shapes and life' by its seminal virtue (More 1646, 7). If this were not the case, these stars would seem to be superfluous. This life-giving light provided to the inert planets by their suns seems to be complemented by two other factors: an inherent activity which God granted to every atom, and an 'ethereal dew' which he scattered all throughout the universe at creation (More 1646, 4, 13).

The poem also contains verses expressing cosmological prophecies in pluralistic terms. It talks of countless other planetary worlds with plants, beasts and men, each beginning with an Adam and ending in conflagration. It suggests that new stars are in fact some of these planets which have met their fiery end, and that comets are perhaps dead planets cast off into the ether (More 1646, 21–22). Nevertheless, the infinite cycle of life continues when these orbs of purged ashes drink in the heavenly dew and 'roscid Manna', resulting in the production of new, prelapsarian life (More 1646, 26). Henry More is an example of how pluralism began to take over the teleo-

[29] See Lovejoy 1948, 125, 167; Koyré 1957, 125–55; Dick 1982, 50–53, 117–18. On Henry More's philosophy and his role in seventeenth-century science, see the various contributions in Hutton 1990 and 2017.

logical function of astrology, attributing meaning to celestial phenomena with drastically new consequences. More's own aversion to astrology will be examined in the next chapter. His substitute in regard to the process of generation was a spermatic 'spirit of nature' or 'world soul'—a universal and general contributor of form and vitality far different and purposefully opposed to the clockwork style of planetary influence of the likes of Vanini. This nominal agent of the immaterial God constituted a Stoic/Platonic plenum, in which each world sits like 'a knot in Psyches garment tide' (More 1646, 4).

A decade later, the French physician and chemist Pierre Borel (1620–1671) wrote his *Discours nouveau prouvant la pluralité des mondes* ('A New Discourse Proving the Plurality of Worlds', 1657), a popular work, translated into English the year after its release, which some have described as a middle-point between the old and new cosmologies (Carré 1974; Dick 1982, 118–20). Borel repeated, as was now usual, the argument that if the earth is a 'star' then the other stars may well be inhabited earths, and he criticised those who thought that the infinite number of celestial bodies had been created for the sake of this earth (Borel 1657, 12). The plurality of worlds may have been a theme widely discussed in France at this time, as it was one of the topics included in the conferences of the *Bureau d'adresse*, a weekly public debate on miscellaneous topics of interest established in 1633 by Théophraste Renaudot (1586–1653).[30] One of the objections raised in this particular meeting was taken on directly by Borel:

> In the second volume of the Conferences *du Bureau d'adresse* someone also opposes that if the stars were inhabited they would need other stars to influence them, and other heavens, so on to infinity; to which I respond that I am not powerfully persuaded that the stars are useful to us, except the sun and the moon. It may be that these stars communicate and serve one another mutually, and therefore there is no need for an infinity of heavens.[31]

The objection is indicative of how many still perceived the entire system of heavenly bodies to be necessary for the functioning of a habitable earth. Borel, on the other hand, saw the generative virtues of the stars as another reason to believe that they are inhabited. If we see causes in the heavens then we should expect to find effects, because we should grant that the stars can act within themselves as well as on distant objects (Borel 1657, 59, 1658, 161–62).

To better explain the operation of such stellar influences, Borel gave the example, unsurprisingly, of the lodestone. A magnet projects its virtue outwards to a certain distance, and the earth, being itself a large magnet, does the same up to the

---

[30] The topic to be debated was in fact the spots on the sun and moon, but the inhabitation of these bodies becomes a central point of discussion: Renaudot 1655, 697–712. The proceedings were later translated into English: Havers 1664, 536–41.

[31] Borel 1657, 48–49: 'Quelqu'un oppose aussi dans le second tome des conferences du Bureau d'adresse que s'il y avoit d'astres habitez, il faudroit d'autres astres pour y influer, et d'autres cieux à l'infiny, à quoy ie responds que ie ne me persuade pas puissamment que les Estoiles nous soyent utiles, excepté le Soleil & la Lune, il peut estre que ces astres se communiquent & servent les uns aux autres mutuellement, & par avoir il n'est pas besoin d'une infinite de cieux.' The English translator unfortunately mistranslates this passage as saying that the stars *cannot* influence each other mutually. See Borel 1658, 131–32.

moon. The other stars likewise extend their virtues and spheres of attraction up to a certain circumference (Borel 1657, 34, 1658, 91). These influences, mutually exchanged, helped Borel prove that an otherwise seemingly unconnected assortment of celestial bodies does in fact unite in *amour mutuelle* to create the harmony of the world. The ethereal air is the ocean which both separates and joins these celestial islands, as it 'equally receives the influences and operations of every Globe, and communicates with great speed to every one those of all the others'.[32] It would not be reasonable, however, to assume that any one of these islands lies uninhabited. We must at least admit, he thought, that they have some plants growing there. But if there are plants, they must be for the use of animals, and animals we know to have been made for the use of man. The whole world, he agreed, was made for the use of man, but rather than having the celestial bodies made solely to influence the earth, he expanded the dominion of mankind to include the infinite number of habitable bodies in the universe (Borel 1657, 68, 1658, 190–92).

Borel's work was as much humanistic as it was philosophical. As well as citing modern authors such as Francis Bacon, Gilbert, Galileo and Descartes, he discussed Cicero's *Dream of Scipio* and ended the work with a long extract from the *Zodiacus vitae* (1536) of Marcellus Palingenius. He cast a wide net for support for his thesis of a plurality of worlds, and the range of implications he discussed was equally broad, including the soteriological status of the humans on other planets and the location of heaven and hell (the former he placed in the sun and the latter in some as yet unidentified star). In Borel's vision, the human population of the earth is no longer at the centre of the universe, but it remains part of the most important species; one island nation among many in the universal air, all connected by a harmony of mutually exchanged influences. The cosmological visions of both More and Borel use the benefits of the sun and moon as a standard model that is repeated in other solar and planetary systems. While More did invoke influence across larger distances, the underlying concepts of harmony and sympathy were robbed of specificity and manipulability, taking on instead more general, aesthetic meanings. Borel did not discuss explicitly astrological influences, but it should be noted that he was deeply interested and involved in the theory and practice of chymistry, which had intrinsic links to astrology and the practical applications of sympathetic powers.[33]

A particularly interesting symbiosis of astrological and pluralist ideas can be found in Otto von Guericke (1602–1686), whose cosmology, as described in his *Experimenta nova (ut vocantur) Magdeburgica* ('The New (so-called) Magdeburg Experiments', 1672), is in many ways reminiscent of Gilbert.[34] Indeed, at one point he paraphrased entire passages straight out of Gilbert's *De mundo*, which by then

[32] Borel 1657, 13: '...toutesfois ce corps sprirituel de l'air recoit esgalement les influences & emanations de chaque globe, & communique tres-promptement à chacun celles de tous les autres.' Borel 1658, 32–33.

[33] See Principe 2013.

[34] Von Guericke 1672. This has recently been translated into English: von Guericke 1994. References are to the English version unless otherwise specified. On von Guericke, see Schimank 1967; Dick 1982, 116; Knobloch 2003.

had appeared in print. He was nevertheless also quite eclectic in his use of literary sources, citing Jesuit authorities such as Giovanni Battista Riccioli (1598–1671) and Athanasius Kircher (1602–1680), and the anti-Copernican Anton Maria Schyrleus of Rheita (1604–1660) to support the fact that the planets are elemental and corruptible bodies like the earth (von Guericke 1994, 40–41). He also quoted Descartes' *Principia philosophiae* ('Principles of Philosophy', 1644) to the like effect, that terrestrial and celestial matter are identical. In a passage lifted straight out of Lansberge, he described the sun as a king on a throne, providing the planets with motion, light and a vital force without which all attempts at generation would be futile (von Guericke 1994, 31–32).[35]

What aligns him with Gilbert, however, is his use of small-scale experiments upon which to build cosmological theories. He is most famous in the history of science for his success in creating a vacuum pump, and he was one of the first since Gilbert to advocate the existence of an interplanetary void. He explained the opinion of the Peripatetics on nature's abhorrence of a vacuum, and how a completely full heaven allows virtues to flow from the upmost bodies through the intermediary ones into the lower (von Guericke 1994, 87). We have already seen how the newer cosmologies still relied on an ethereal plenum to convey influences. Von Guericke's own theory was quite different. He did not call the general container of the earth-like bodies ether, nor exactly vacuum, but 'space' (*spatio*); a space which is full when such a body with its emanations is present, and empty when one is not (von Guericke 1994, 88–89). One such emanation is air, which he described, in Gilbertian style, not as an element but as an effluvium of the solid and liquid parts of the earth (von Guericke 1994, 111). This effluvium is a corporeal virtue which flows out in to the surrounding space up to a prescribed limit, at which point pure space begins. No corporeal body should be thought to exist beyond that point. Von Guericke, like Gilbert, believed that such matter, fluid or otherwise, would hinder rather than help the motion and influences of the celestial globes (von Guericke 1994, 131).

The air is one corporeal virtue of the earth, but there are other such 'mundane' virtues common to all earth-like bodies, such as the planets and their moons, and also the sun. These virtues are either corporeal or incorporeal, and they extend themselves out into a sphere of virtue or activity which rarefies and weakens with distance. Corporeal virtues in general are effluvia from solid bodies which cannot pass through other solid bodies, while incorporeal virtues can (von Guericke 1994, 193). He talked of these latter as effluvia as well, but had earlier made it clear that they are completely without body. An incorporeal virtue does not fill up space and only 'exists' as long as a body which is susceptible to its influence is present within its sphere of activity (von Guericke 1994, 133–34). He identified six incorporeal virtues of the earth (impulsive, conserving, directing, turning, sounding and heating), two of the sun (light and colour) and one of the moon (chilling). There are undoubtedly, he thought, more virtues directed towards the earth from the planets, which astronomers call 'influences' (von Guericke 1994, 193–94).

---

[35] Compare von Guericke 1672, 20, with Lansberge 1630, 38.

Again, these larger cosmological theories were based on experimental observations, and with regards to spheres of virtue his main guides were magnetism and electricity. For example, he discovered that a rubbed sulphur globe attracted some light objects, but arbitrarily repelled others, which seemed to him to explain 'how the influences of the planets vary according to the differences in their forms and their proximity to one another' (von Guericke 1994, 207). This endorsement of varied and individual celestial planetary influence was even extended to include astrological aspects:

> Moreover, just as we are aware that many virtues emanate and flow from this body of the earth, diffusing themselves far and wide in surrounding space, so also must we assume that the rest of the mundane bodies or planets individually release different virtues consonant with their own natures and qualities, and that these touch and affect one another according to the nearness or farness of their positions, and their different situations (as when they look toward one another in opposition, conjunction, quadrature, sextile, etc.). Astronomers are wont to call these virtues 'influences', which are either good or evil depending upon the different times and the different effects of these mundane bodies upon one another, and in any one body the Astronomers allege many effects.[36]

Although he accepted the theory of aspects in principle, however, and knew that the influences of the sun and the moon are so indispensable that the functioning of life on earth depends on them, he believed that the planetary influences are not so individual as to allow certainty in predictions, and he quoted such authorities as Marin Mersenne (1588–1648) and Riccioli on the inaccuracy of judicial astrology (von Guericke 1994, 232). Also counting against the astrologers was the presence of hitherto unknown sources of influence, such as sunspots, which von Guericke believed to be small planets orbiting very close to the sun (von Guericke 1994, 34).

While there is this largely beneficial interplay of virtues and influences between the globes—the sun even feeds, in Stoic style, off the air from the planets—there is also quite a lot each globe can do for itself (von Guericke 1994, 321). Von Guericke believed (with similar Stoic bent) that the earth has a soul and that each celestial body similarly has a life and soul of its own (von Guericke 1994, 193). This is where he copied straight out of Gilbert's *De mundo*, even using the same quotes from Virgil ('the mind moves the mass') and Aristotle ('the soul is the activity of an organic body'), and stating that the *actus*, diffused through the entire body of the earth, is ultimately responsible for all birth and growth.[37] Again, he echoed Gilbert

---

[36] Von Guericke 1672, 150: 'Porro sicut experimur et percipimus, multas ex hoc Terreno corpore egredi vel effluere virtutes, sese in spatium circumstans longe lateque diffundentes; sic quoque de reliquis corporibus mundanis sive Planetis non dubitandum, eos singulos pro cuiusvis natura et qualitate, diversas exspirare virtutes, et illis sese invicem pro diversa eorum, vel proximiore vel remotiore statione, ut et diverso situ (quando scilicet sese adspiciunt ex Oppositione vel Conjunctione, vel Quadrato vel Sextili, etc.) tangere aut afficere. Quas virtutes Astronomi communiter Influentias vocant, quae pro diversis temporibus ut et affectibus horum mundanorum corporum erga se invicem, vel bonae vel malae sunt, et in quolibet corpore plurimos effectus causantur.' This is an altered version of the translation in von Guericke 1994, 231–32.

[37] Compare von Guericke 1672, 156: 'Anima dicitur actus corporis organici: Telluri non solum actus est insitus per totam molem, sed etiam adnascentium, excrescentium, supervenientium etc. entelechice ab illo pendent. Organa si desideres, ecce polos duos…' with Gilbert 1651, 125:

that the earth's soul is responsible for its motion, turning itself so that all things living on its surface can receive the light from the sun—the difference being that von Guericke explicitly mentioned the earth's annual motion.[38] His familiarity with both Gilbert's *De magnete* and *De mundo* could help explain his experimental interest not only in magnetism and electrical attraction but also in the vacuum, a subject much more fully developed in Gilbert's posthumous work. Von Guericke did add something seemingly novel to Gilbert's theory of a living earth. If it is alive, it should grow, and this growth should lead to incrementally slower motion, which could explain the precession of the equinoxes (von Guericke 1672, 157, 1994, 244).

The importance of this planetary soul for the generation of living creatures came into sharp relief when von Guericke discussed the moon. He was one who argued against it being inhabited, or at least, that any life on it must be of a much lower order. He gave several reasons for his position. One was that it is too small in proportion to its distance from the sun to retain enough heat. He believed there was a direct correlation between distance from the sun and planetary size, so much so that he criticised those who estimated Mars to be smaller than the earth (von Guericke 1994, 334). This lack of friction with the solar rays also meant that the moon doesn't have an internal fire (von Guericke 1994, 276–77). Another reason was that the chilling influence of the moon on the earth suggested that that entire body is frigid. One further reason was that the moon doesn't have any animate spirit of its own, which von Guericke thought was demonstrated by its lack of rotational movement. Without such an animate virtue, it has no hope of generating or sustaining animated creatures on its surface (von Guericke 1994, 282–83).

The moon, however, was an exceptional case, with von Guericke populating all the other planets and even the interior of the earth with living creatures (von Guericke 1994, 242). One of the reasons he gave for believing the planets to house animals was that they possess similar motions to the earth, which makes sense once you consider his association of a motive planetary soul with the ability to produce mobile creatures. The other reasons he gave are familiar by now: (a) it is hard to believe that such huge bodies are devoid of life; (b) they don't seem to have been created simply to provide light to the earth; (c) the omnipotence of God is not limited to this one tiny corner of the universe (von Guericke 1994, 338). Unlike Borel, von Guericke was quick to argue that these creatures would not be men. The reason he gave is based on diversity. He stated that 'no animal which is on this sphere of

---

'Anima est actus corporis organici. In tellure non solum actus est insitus per universam molem; sed etiam adnascentium, excrescentium, supervenientium entelechiae ab illo pendent. Organa si desideres, ecce tibi polos…' The mistaken transcription or printing of 'entelechiae' as 'entelechice' in von Guericke causes problems for the modern English translator, who understandably treats it as an adverb and is left with dangling genitives. See von Guericke 1994, 242–43.

[38] Von Guericke 1672, 156: 'Solemque (a quo recreatur et simul in circuitu magno seu annuo circumducitur) observare, ac eius lumen, convertendo sese ob salutem suam et omnium supra suum corpus nascentium et viventium incolumitatem, recipere possit.' Cfr. Gilbert 1651, 126: 'Solem observat, a quo recreatur impelliturque, cuius lumina, convertendo sese ob salutem suam, et omnium supra eius corpus nascentium incolumitatem, recipit per vices et successiones.'

earth also exists on the other planets, since there is a diversity in all things causing everything which is there to differ basically from the bodies we know' (von Guericke 1994, 338). The belief that God delights in variety was quite orthodox philosophically, and it also allowed von Guericke to sidestep theological questions. There may be rational animals on other planets, he admitted, but their natures, intellects and societies will be different, and God will act with them in different ways (von Guericke 1994, 340). In the case of theories about the nature of ET life in this period, it is tempting to think of a 'variety' camp and a 'similarity' camp, but the treatments of this subject are usually not detailed enough to allow such a clean distinction. For example, in von Guericke's case it is uncertain whether this biological variety is on a substantial, functional or simply accidental level.

We have seen thus far in von Guericke a cosmology very similar to that of Gilbert, but there are some important differences. Von Guericke openly adhered to the Copernican system, and he discussed atoms (briefly) and the probability that the stars are other suns. Still, we have an assortment of ensouled globes, separated by empty space, moving themselves so as to preserve habitable conditions on their superficies and surrounding themselves with emanations and spheres of influence. These last are key to understanding how von Guericke maintained a cohesive whole out of a collection of disparate globes in empty space:

> And even if so vast a space is assigned to each one of these bodies by the Copernicans, still this is not to be regarded as a useless vacuum. Indeed it must so please the divine Immensity to place its creatures widely apart for its own greater glory that each may be of an atom's proportion relative to the vast Space surrounding it. And just as it is the nature of all living things in this world always to be seeking greater space, and just as all things which are of the same type have a neighbor and seek to extend themselves up to the point that it itself extends... in the same way also, one can surmise that each of these heavenly bodies extends itself far and wide by means of its exhalations up to the limits of its own kind and expands its virtues between itself and its neighbours. Thus, in accordance with the condition, nature, mass, quality, etc. of each, these bodies function, come together, influence, and look upon one another, for good or ill (von Guericke 1994, 356).

Empty space, yes; but useless vacuum, no. It is occupied immaterially by the spheres of virtue of the planets, within which they act at a distance on other susceptible bodies. In von Guericke's cosmology there are pieces of astronomy, astrology, laboratory physics and biology all playing their part to explain the size, shape, mechanisms and functions of the universe.

Von Guericke's main opponent in his arguments for the existence of empty space was Descartes.[39] While Descartes himself said little about the possibility of ET life and was scornful of astrology, subsequent philosophers more strictly Cartesian than More or Borel explored both subjects in respect to a universe filled with particles and vortices.[40] Antoine Le Grand (1629–1699) wrote in his *Institutio*

[39] He explicitly framed his argument against those who equated extension with matter, which was one of Descartes' fundamental principles: von Guericke 1994, 131. See Descartes' *Principles of Philosophy*, Book I.

[40] Descartes did, however, make an interesting remark to Mersenne in 1632, suggesting that a knowledge of the natural order of the seemingly irregular dispersion of the fixed stars could enable

philosophiae, secundum principia Renati Descartes ('The Teaching of Philosophy According to the Principles of René Descartes', 1672) that heavenly matter undergoes generation, which is a simple 'congruous adaptation' of material parts, and that there may be many earth-like bodies inhabited by diverse animals (Le Grand 1694, 138–39). He also maintained that these far distant parts of the world are united by a mutual relation and a uniting virtue, and this virtue is 'the Subtil or Aethereal Matter, which permeating all Bodies is the cause of this Union and Harmony' (Le Grand 1694, 141). This ethereal matter, spread everywhere like a *pneuma*, allows for a form of celestial influence which reads like Cartesian-flavoured Stoicism: 'For the Stars entertain such a Communication among themselves, as to convey food to one another through their Vortexes, bestowing upon others what goes out by their Ecliptick, and receiving from others what comes in by their Poles' (Le Grand 1694, 158).

Le Grande, like other prominent Cartesian philosophers, was critical of the practice of astrology, and this limited action of particle transference is conventionally Cartesian. Jacques Rohault (1620–1675) included a chapter in his popular *Traité de Physique* ('Treatise on Physics', 1671) advocating the abandonment of astrological pursuits.[41] He described the moon and planets as material and earth-like, but took an agnostic position on whether they were inhabited. At least one Cartesian philosopher, however, the young Claude Gadroys (c. 1642–1678), attempted to justify larger swathes of astrological theory and practice according to Cartesian mechanical principles.[42] In his *Discours sur les influences des astres* ('Discourse on the Influences of the Stars', 1671), he argued that heat and light alone cannot explain the diversity of effects seen within the vortex of the earth, and so some time should be given to the examination of other possible forms of influence.[43] In his physical explanation of the process of influence, he described how the sun strikes the planetary bodies and in doing so elevates small particles. These particles are then pushed out with the rest of the vortex, taking with them the property of their origin (Gadroys 1674, 61–62).[44] Seeing as a void is impossible in nature, each body must receive as much material as they emit, such that all celestial as well as sublunary bodies are constantly sustaining themselves one to another. 'It is a bond by which God has united all these bodies,' he declared, 'without which they would be disturbed in a thousand different ways, and never retain any assured rule.'[45]

---

a priori knowledge of the forms and essences of material bodies. The passage is quoted in Jaki 1978, 39. On Cartesian astrology in general, see Bowden 1974, 197–98; Spink 2017, 111–13.

[41] The work went through multiple French editions, as did the Latin translation by Samuel Clarke and John Clarke's English translation from the Latin. See Dick 1982, 113–16; Rutkin 2006, 556.

[42] Gadroy's astrological theories were discussed by Lynn Thorndike (Thorndike 1923–1958, VII, 559–62, 680–82). See also Bowden 1974, 198–202; Spink 2017, 118–29. Gadroys gets a brief mention in Dick 1982, 116.

[43] References are to the second edition, Gadroys 1674, 7–8.

[44] This is similar to the mechanistic conception of celestial influence theorised by Isaac Beeckman (1588–1637). See Vermij 2016, 306.

[45] Gadroys 1674, 61: 'C'est un lien dont Dieu a uni tous ces corps, sans lequel ils se remuëroient en mille différentes maniéres, et ne garderoient jamais aucune regle asseurée.'

Most interesting for our purposes is that within this work on celestial influences is a chapter on the nature of the planets which presents a balanced discussion on the possibility that these bodies are inhabited. Gadroys believed that the best way to understand the nature of the planets is to consider this earth of ours and the bodies which compose it (Gadroys 1674, 43). Observations and terrestrial analogies prove that they are round, solid, opaque and rough bodies, with certain protrusions and possibly even water (Gadroys 1674, 43–46). All these factors led him to ruminate on the subject of inhabitability (Gadroys 1674, 46–50). He thought it reasonable to suspect that the combination of moisture and sunlight would lead to the production of plant life on the planets, but hesitated to conclude that they would be inhabited by animals and men as well. Displaying a healthy Cartesian distaste for teleological thinking, Gadroys offered rebukes to philosophers on both sides of the argument who use the dictum that God does nothing in vain. We should not assume that we are the sole purpose of creation, but nor should we assume that God cannot take glory from the non-living parts of his creation. 'That is why we must suspend our judgment', he concluded, 'and believe that it is as probable that there are men in the Planets, as it is that there are none.'[46] This agnostic declaration, similar to that of Rohault, is then followed by this summation: 'Though we have remarked that all Planets are alike in being solid, hard, opaque, fertile, and habitable, it may nevertheless be said that they are all different from one another.'[47] In order to preserve a functioning astrological system, Gadroys—similarly to Gilbert—balanced the essential similarity of the planetary bodies with an assertion of the individual differences between them.

As mentioned in an earlier chapter (see Sect. 2.4), it should not surprise us that a work on astrology discusses the possible habitability of the planets, concerned as astrology is with the nature and physical composition of the celestial bodies. Another example of this is the English translation of Manilius' *Astronomica* made by the poet Sir Edward Sherburne (1616–1702). In this quite impressive edition, copious notes and appendices accompany the text of the astrological poem itself, giving explanation and context as well as informing the reader of more modern opinions on the astronomical subjects. The references it gives to modern authorities are quite thorough, from Galileo, Kepler, Gassendi, von Guericke, to the Jesuits and the Cartesians, usually with specific chapter references. Sherburne informed the reader of the modern opinion that each star is the centre of a planetary system, and that our sun is a body mixed of liquid fire and 'asbestinous' solid parts from which flows a 'panspermatick virtue' (Sherburne 1675, 156, 165).

The section of the appendix dealing with the moon concludes that it is a terraqueous globe like the earth, which has been placed at such a convenient proximity that it may seemingly intercept and temper the light and influence of the celestial bodies,

[46] Gadroys 1674, 49: 'C'est pourquoi il faut suspendre notre Jugement, et croire qu'il est autant probable qu'il y ait des hommes dans les Planétes, comme il est probable qu'il n'y en a point.'

[47] Gadroys 1674, 49–50: 'Quoique nous aions remarqué que toutes les Planétes se ressemblent, en ce qu'elles sont solides, dures, opaques, fertiles et habitables, on peut néanmoins dire qu'elles different toutes entr'elles.'

especially the sun, before reflecting them off its uneven surface towards the earth 'with less incommodity' (Sherburne 1675, 169). Sherburne warned the reader not to assume that the stones, lakes and other such things on the moon are made of the same material as ours; in fact they are quite different and incomprehensible to us. Yet its usefulness to us and the alien nature of its material does not prevent it from being inhabited. Indeed, Sherburne believed that the recent telescopic observations of 'mountains, vallies, woods, lakes, seas and rivers' simply confirmed a fact known to the ancients, from Xenophanes to Macrobius (Sherburne 1675, 179). What kind of creatures these Lunarians may be is unknown, but their existence is almost certain, seeing that nature has provided the moon with everything it needs to sustain life. It should not be thought, argued Sherburne, 'that these Advantages and Benefits should be conferred by Nature for no Use or End; or that the Moon should only be made to reflect the Sun's Light to us' (Sherburne 1675, 179).

This mixture of astrological and pluralist ideas might seem like a clash of the old and the new, but for Sherburne both ideas were equally ancient and equally modern. He relied on the authority of Kircher to explain the nature of Mars, with its solid substances akin to sulphur and arsenic 'evapourating malignant and destructive qualities' (Sherburne 1675, 182). The work he cited is Kircher's monumental *Itinerarium exstaticum* (1656), which was published more than two decades after Kepler's *Somnium*. There were Jesuits and figures like Gadroys who were applying new astronomical discoveries to astrology just as there were writers like More and Borel directing astronomy towards pluralism. For a humanist like Sherburne each project had an ancient tradition, and there was no reason to consider them mutually exclusive.

There were some who gave more of the credit for pluralism to modern astronomers. Robert Wittie (1613–1684) in his *Ouranoskopia* (1681) talked of the inhabited moon and planets as things unthought of by the ancients (Wittie 1681, 26). He considered that great and unprecedented steps had been made in the last century, and that with so many 'Excellent Heads' at the Royal Society more could be expected in time (Wittie 1681, 35–36). He believed it arrogant to think that the planets and stars have been made for our use, and was doubtful that judicial astrology could be compatible with Christianity, and yet still celestial influence was a major part of his cosmological vision (Wittie 1681, 30–31, 57). Indeed, influence was part of his argument for inhabited planets:

> We plainly discern them to be dark Bodies like this Globe of the Earth, and to have a continual Succession of Day and Night, and Moons that surround them, that give Light by Night to them; and we may probably guess that there are other Influences that must according to the Course of Nature flow from them, and operate upon their several Planets, as the Moon upon this Globe of Earth and Sea; but what signifie all these, if there be no Inhabitants, or rational Beings in them? (Wittie 1681, 29).

If God, as he argued, uses the same methods of providence everywhere in the universe, then there should be no doubt that those other satellites are performing the same life-promoting functions for their planets as our moon does for us.

Wittie talked of the 'Machine of the World', a metaphor appropriated by the mechanical philosophers but originally applied to the world of top-down astrological causation. The significance of the metaphor to Wittie related not so much to the regularity or transfer of motion, but to the dependence of each part on all the others, centred on influence:

> And now let every wise Man and good Christian sit down a while, and consider what has been said in the foregoing Discourse concerning this great Machine of the World, so made and ordered by the Wisdom and Power of God, consisting of so many, and so great stupendous parts, at such immeasurable Distance from one another; yet with such subserviency towards one another, as that they cannot subsist, nor continue to perform the Ends of their Creation without the Help and Influence of each upon other (Wittie 1681, 40).

This vision, which we have seen now in many forms, was a direct product of the confluence of the astrological paradigm with a belief in a plurality of earth-like worlds. The increased size of the universe in the new astronomy strengthened already present suspicions that God had not created all the innumerable celestial bodies solely for the benefit of the earth, and this was met with other theories and observations which suggested that many if not all of these bodies were habitable. Yet this was not enough to dispel the ancient and strongly held belief in a universe defined by influential relationships. The symbiosis of astrological and pluralist themes arose out of its use as a rationale for celestial motion and the reluctance to abandon the harmonious interconnected unity of the Aristotelian cosmos, as well as out of the shared intellectual and to some extent professional context of the two ideas. Our discussion will now begin to turn to how these two paradigms became delineated, philosophically opposed and contextually separated. To that end, the rest of this chapter will examine the debate over astrological and pluralist possibilities between differing schools of early mechanism.

5.3 Influence and Inhabitation in Early Mechanism: Digby, White and Hobbes

On the surface, the mechanical philosophy seems to be favourably inclined to pluralism, with its general commitment to heliocentrism, its claim to describe all phenomena, including life, in terms of matter and motion, and its abandonment of the ontological divide between terrestrial and celestial substance. If the ingredients and processes which lead to the development of living creatures are not unique to the earth, then the generation of life on other celestial bodies may follow as a logical consequence. By the same token, its aversion to 'occult' processes (in this sense meaning processes not involving, or involving more than, matter and motion) and its denial of any special dignity to the heavens would seem to make it incompatible with astrology. These are, of course, gross simplifications. Under the loose categorisation of mechanism lay any number of distinct cosmologies, as we have seen already, and commitment to celestial influence and/or inhabitation did not only concern physics, but also biology, theology, morality, epistemology and more. This complexity is demonstrated in the natural philosophies of Sir Kenelm Digby

(1603–1665), Thomas White (1593–1676) and Thomas Hobbes (1588–1679). All three figures were in some way connected to the intellectual circle of Marin Mersenne in Paris, which also included the likes of Descartes and Gassendi. The philosophical agreements and disagreements of these English exiles demonstrate the intricacy of the interactions between atomism, religion, astronomy and the life sciences.

The three main works to be considered are Digby's *Two Treatises* (1644), White's *De mundo dialogi tres* (1642) and Hobbes' refutation of White's *De mundo*, composed in 1643.[48] Although Digby's work was published after the composition of the other two works, White's references to it indicate that a draft manuscript existed prior to 1642. Indeed, the two works deliberately complement each other, with Digby providing the physical and metaphysical foundations for White's broader cosmological speculations. The two men were reforming Catholics and reforming Aristotelians, and their works are primarily concerned with reconciling aspects of the new astronomy and atomism with core tenets of Catholicism and Aristotelian philosophy. Before dealing with White's cosmology and Hobbes' nearly point-by-point rebuttal, we will begin with the *opus maius* of Digby, a figure who, while popular and well-renowned in his time, remains relatively un-feted in the history of science.

5.3.1 Kenelm Digby

The first and much longer treatise in Digby's major philosophical work treats the nature of bodies, while the second treats the immortal soul.[49] The attested aim of the work is to demonstrate the inherent limits in any physical philosophy so as to prove the necessity of a belief in an immaterial, immortal soul. His philosophy, as others have noted, was a development of the *minima naturalia* tradition, rather than Epicurean atomism; he denied the existence of indivisibles or the void, and defined atoms as 'the least sort of natural body' (Digby 1645, 14, 25–27, 48; Clericuzio 2000, 82; Henry 2010, 70, n. 35). The physical treatise begins with the axiom, taken from Descartes, that body is the same as extension. Digby believed that the only differentiating factor in matter was a scale of rarity and density. From this foundation, he believed he could show that 'all Physicall things and naturall changes do proceed out of the constitution of rare and dense bodies' (Digby 1645, 31). Following a straight Aristotelian definition, he explained that a body is rare when its quantity is more while its substance is less, and is dense when the substance is more, the

[48] References to Digby's work are to the second edition: Digby 1645. Hobbes' untitled Latin criticism of White's *De mundo* has only been published in modern times. Latin references are to Hobbes 1973. English translations are from Hobbes 1976. On the dating of Hobbes' critique, and for a discussion of the shared intellectual context of all three philosophers, see the introduction to Hobbes 1973, 1–46.

[49] For an introduction to Digby's context and philosophical agenda, see Henry 2010.

quantity less (Digby 1645, 28). Heat is explained as a property of rarity; cold of density. The action of gravity then produces the qualities of moist and dry. A body is dry if its rarity or density is in greater proportion than the gravity which works upon it; moist if less (Digby 1645, 33–35). Having thus explained the foundation of the four prime qualities, Digby could easily combine them to explain the origin of the four Aristotelian elements, although it was patently un-Aristotelian to use gravity prior, rather than subsequent, to the distinction of the elements (Digby 1645, 37–38). Even less Aristotelian was his assertion that 'out of all we said of these foure Elements, it is manifest there cannot be a fifth' (Digby 1645, 39).

Out of the four elements, one in particular, fire, plays a particularly important role in physical change. This is because Digby considered light to be nothing but fire, spreading far and wide, unmixed with other 'grosse bodies' (Digby 1645, 56). The interaction of fire with the other elements determines the temper, consistence and generation of all mixed bodies, and thus the fire of the light from the sun is 'a universal action in the generation and corruption of bodies' (Digby 1645, 113, 178). The light of the sun is even made responsible for gravity, which, as noted, is no longer a natural inclination of the heavier elements. Digby suggested that the light from the sun strikes the earth and rebounds, breaking off and carrying with it small dense particles. As these particles are lifted, other less dense particles rush in to take their place, moving in a direction which bisects the angle of reflection, and hence always perpendicular to the surface. Once the sun goes down, the denser particles which had been lifted up, 'finding themselves in a smart descending stream', move back down, displacing the rarer bodies below them (Digby 1645, 95–97). Thus all places within reach of the sun's action will experience this perpetual motion of the air which results in gravity.[50]

Digby employed similar mechanical explanations in the generation of plants.[51] He asked the reader to imagine a buried conglomeration of mixed bodies with an excess of fire but very little moisture. The arrival of water to this mixture permits the free action of the fire, which causes the body to swell and grow. Endeavouring not to resort to the action of pre-existing form, he takes us through a step-by-step process of development, based on the nature of the mixed bodies and the actions of internal fire and external sunshine, resulting in a fully-grown plant with roots, branches and fruit. Juice, proceeding from the roots and being passed, strained and concocted throughout the whole plant, condenses in the centre of the fruit 'like a tincture extracted out of the whole plant'. There it dries and becomes a seed, which falls into the earth and awaits the arrival of moisture to start the operation all over again (Digby 1645, 260–62). Digby then described the natural process by which a plant may develop systems and organs granting it sense and movement, thus becoming a sensitive animal (Digby 1645, 262–65). While a seed or egg might ensure the continuity of species, Digby rejected the notion that it is necessary for the generation of plants and animals. As the start of his description of plant generation would

[50] Digby's involvement of air in a theory of gravity leads him, like Gilbert, to deny its action below the surface of the earth: Digby 1645, 110.

[51] For an introduction to mechanism and biology, see Roger 1986.

suggest, he believed that spontaneous generation from the natural action of heat and moisture on suitable mixed bodies was both possible and not uncommon (Digby 1645, 268–70). Indeed, plant and animal life resulting only, like all other physical things and actions, from the mixture of rare and dense bodies, he argued that it is not only possible spontaneously, but also, daringly, that it is 'easie to be effected' (Digby 1645, 270).

Digby thus described processes from gravity to generation in terms of the action of fire (including sunlight) on and within substances mixed from rare and dense bodies. Heat and cold were both active qualities in his estimation, but there were other, more specific actions and properties to explain, including magnetism. Rather than going into magnetism specifically, it will suffice to quote some more general arguments of Digby's that pertain to many similarly 'occult' activities:

> … there is not a body in the world, but hath about itself an orbe of emanations of the same nature which that body is of. Within the compasse of which orbe, when any other body cometh that receiveth an immutation by the little atomes whereof that orbe is composed, the advenient body seemeth to be affected and as it were replenished with the qualities of the body from whence they issue (Digby 1645, 172).

The major difference between this description and that of, say, Gilbert, is Digby's invocation of atoms and his denial of immaterial emanations. Regardless of this denial of effused forms, Digby could still rescue a physical explanation for sympathy that accounted for action at a distance:

> But withall we must note that it is not our intention to say, but that it may in some circumstances happen that some particular action or effect may be wrought in a remote part or body, which shal not be the same in the intermediate body that lieth between the agent and the patient, & that conveyeth the agents working atomes to the others body. … The reason of which is manifest to be the divers dispositions of the different subjects in regard of the Agent: and therefore it is no wonder that divers effects should be produced according to those divers dispositions (Digby 1645, 173–74).

Such a broad supposition worked to explain the effect of the lodestone on iron, but it also explained the action of the substance for which Digby was most famous: the weapon salve. His discourse on this powder of sympathy (Digby 1658), which was thought to be able to treat wounds by being applied to the offending weapon or bloodied bandage, went through many editions and translations (Thomas 1971, 225). Much of it was a summary of the relevant parts of his *Treatise on Bodies* with the addition of anecdotes intended to prove the veracity of his theories, as well as long digressions on the effects of sympathy in human beings themselves.[52]

The salve is simply powder of vitriol that is dissolved in water and has the bandage soaked in it. Sunlight strikes the bloodied bandage, carrying off small atoms of blood with the vitriol joined to it, diffusing the atoms into an orb. The wound meanwhile is exhaling fiery spirits, and consequently drawing in the surrounding air. The bloody atoms from the bandage then find themselves attracted back into the wound, where they find their 'proper source, and originall root', at the same time delivering their attached medication (Digby 1658, 133–35). The entire process, then, amounts

[52] On this last, see Lobis 2011.

to a targeted delivery system for a common medication. Digby went on to discuss the nature and effects of vitriol, which he described as one of nature's most excellent bodies. 'The *Chymists* do assure us', he wrote, 'that it is no other then a corporification of the universal spirit which animates and perfects all that hath existence in this sublunary World' (Digby 1658, 142). Digby had a long interest in alchemy and astrology, and this universal spirit was central to both pursuits.[53] However, this was no Platonic 'Soul of the World'—it still remained within the realm of physics and mechanical explanation.

Digby refrained from much discussion of the universal spirit in his discourse on the weapon salve but gave a slightly longer definition in his *Discourse Concerning the Vegetation of Plants* (1661). Here he was discussing a nitrous salt which he considered to be responsible for the generative powers of rainwater and dew:

> Now in this Salt are enclosed the Seminary vertues of all things. For, what is it, but a pure extract drawn by the Suns-beames from all the bodies that he darteth his Rayes upon, and sublimed up to such a height of place as leaveth all feculence behind it, and is there in that exalted Region of the Limbeck baked and incorporated with those very beames themselves which refined this extract out of its drossie Oar? Therefore I wonder not to see any sort of hearbe grow upon the highest Towers, where it is certain no man ever came to sow that Plant. And the Loadstone or Magnes of a like substance (though nothing near so pure) that is in the Earth, the creeping toad there, sucketh and pulleth down this flying Dragon to it; and both of them do become one body. And thus you see plainly and familiarly explicated the great Aphorisme of the Smaragdine Table; That what is above, is like what is below. The Sun is the Father, the Moon is the Mother; the Earth is the Matrix wherein this product is hatched; and the Aire conveyed it thither. This Universall Spirit then being Homogeneall to all things, and being in effect the Spirit of Life, not onely to Plants, but to Animals also: were it not worth the labour to render it as usefull to mens bodyes, as to the reparations of Plants? (Digby 1661, 68–70).

The sun is performing the same task as the juices in a plant, or the alchemist in his lab, extracting the essence from everything within reach of its beams, and then distilling these traces into a fine yet material spirit. This spirit is then re-solidified for use as either a universal stimulant or a panacea, depending on the situation.[54]

So much of Digby's philosophy rests upon the action of fire in the form of the sunlight. It is the universal cause of generation and corruption, it is the cause of gravity, and by rebounding off objects big and small it creates orbs of atomic emanation. Its action is universal. It tears off cold and moist particles from the moon and carries them by reflection to the earth, just as it carries vitriol to the wound. In the *Two Treatises* is a short passage which suggests that such universal action is accompanied in Digby's mind by some idea of ET life. Section 21 of Chapter 14 in the 'Treatise on Bodies' is marginally headed by the statement that 'in the planets and starrs there is a like variety of mixed bodies caused by light as here upon earth'. The main text reads:

---

[53] See Dobbs 1971, 1973, 1974.

[54] Boyle was one later philosopher who was interested in this theory of Digby's. See his *Suspicions about some hidden qualities in the air* (1674), in Boyle 1999, VIII, 121–142. See also Henry 1986, 344.

> ...wheresoever there is any variety of active and passive bodies; there mixed bodies likewise must reside of the same kinds, and be indued with qualities of the like natures, as those we have treated of; though peradventure such as are in other places of the world remote from us, may be in a degree far different from ours (Digby 1645, 158).

Given there can't be more than four elements, and presuming the universal action of light upon those elements, we can postulate the same kinds of generations and corruptions in the other planets, including the continuing spontaneous generation of living creatures. Digby, however, did not expand on this point.[55] He kept to the limits of his work which was concerned entirely with familiar terrestrial objects and phenomena. While keeping his feet on solid ground, he entreated the reader to refer to the work of others who had sailed the ocean of higher cosmological speculation and brought back solid reports: 'Which surely our learned countreyman, and my best and most honoured friend... hath both profoundly, and acutely, and in every regard judiciously performed in his Dialogues of the world' (Digby 1645, 180).

5.3.2 *Thomas White*

Thomas White was Digby's friend, mentor, fellow Catholic and confederate in the attempt to create a new synthesis of Aristotelianism, Christianity and mechanical philosophy.[56] He also wrote dozens of theological works under his alias Thomas Blackloe, which gave rise to the Blackloist movement in English Catholicism.[57] Here, however, we will focus on his major work of natural philosophy: his *De mundo dialogi tres* (White 1642). The cosmology therein is an innovative attempt to reconcile Copernicus and Galileo with Aristotle and Christianity, all within a largely mechanical framework. White rejected, for example, a universe of infinite extent, yet believed it to be big enough that the entire annual orbit of the earth could be considered a centre point.[58] The cause White gave for the motion of the earth was particularly creative. He argued that the wind drives the ocean tides, and this movement of the ocean pushes the earth like weighted bags on a perpetual motion wheel. The first problem was that the prevailing winds move in the opposite direction to the rotation of the earth, but White solved this by arguing that the surface water, which is driven by the wind, pushes deeper water back the other way. It is this deeper water which is in contact with the earth, and drives it from west to east.[59] This was, of course, an inversion of Galileo's hypothesis, which had argued that the tides were caused by the rotation of the earth.

---

[55] An interesting point to raise here is that Pierre Borel later dedicated his *Discours* to Digby. The two men were apparently part of the same circle of 'chymists' working in Paris in the 1650s and 1660s. See Principe 2013, 24.

[56] See Henry 1982.

[57] See Henry 2010, 46–51.

[58] See Southgate 1993, 94–95.

[59] See Southgate 1993, 98–99.

These are some of the higher speculations which Digby left out of his *Two Treatises*, but they are soundly built on his principles. White took up the issue of cosmology where Digby had left it, discussing the similarity of substance and action in the heavens. There are three interlocutors in the dialogues: Ereunius gives the author's opinion, Andabata is the antagonist and Asphalius is the notionally neutral third party. Andabata is sceptical of Ereunius' claim that things which are born and die in the moon and the other stars are similar to those here on earth. He suggests that Ereunius limit his argument to the claim that the heavens are not free from corruptibility. After all, all our knowledge depends on sensory experience, and as we cannot have any particular experiences of the situation on the moon, we cannot extend the argument further (White 1642, 55–56). Ereunius replies:

> Not so fast. You know what they say about the hasty dog. If you had but read a certain little golden book (which I have avidly considered) on the Immortality of the Soul, you would know otherwise. So that you may understand regardless, I will put forth the whole argument of this book, which can be ascended like a ladder up to the summit of heaven, and to the knowledge of those things which happen there, of whatever kind. This most ingenious man, seeing that the greatest power is positioned in that place which he has shown to be the spiritual soul of man, it follows that it [the soul] is immortal. Seeing also that something immaterial cannot be known of itself, except by way of separating body from spirit, he judged he should aim at that very place, to explain what can be attained by body, and what place is left to spirits.[60]

This 'little golden book' is the *Two Treatises* of Digby. While it is referred to here by the name of the second treatise, the argument is based exclusively on the physics of the first. Ereunius continues:

> What, therefore, did he come up with? That body is the same as extension is a fact appreciated by children and the elderly alike. From this beginning, he delineates quantity through rarity and density, in a way dividing by primary differences. But where there is rare and dense, being placed relative to a part of the universe, from necessity gravity and levity are born. Where these are present the four primary qualities, hot and cold, wet and dry, cannot be absent. And from these foundations we have the necessary reason of the Elements, each of which can be made by putting two of these Aristotelian qualities in order. But the elements cannot be observed in any great mass, or in their own estates. The particles, however, being *minima*, are continually mixed together. Hence having affixed the common reasons of mixtures, he explains the causes and origin of various motions, extending himself up to plants and animals. In every place he shows things which are produced, except man, to be of such a kind, that they follow from a necessary joining together from the organs of nature and the complexions of rare and dense, like threads being sent out and interweaved from a spider's cave. After which he also illuminates the activities of the soul, and teaches that they

[60] White 1642, 56–57: 'Nihil tam propere; nosti enim diverbium, quod de Cane festina solet iaci; Tu vero, si libellum quemdam aureum (quem ego avide pervolutavi) de Immortalitate Animae legisses, aliud sentires. Et ut hoc saltem videas, quibus scalis Arx Coeli superetur, et ad notitiam eorum quae ibi geruntur, qualemcumque ascendatur; totius libri argumentum tibi expono. Videns itaque vir ingeniosissimus in eo maximam vim positam, ut animam hominis spiritualem esse convinceret, quo immortalem esse consequeretur; neque posse immaterialitatem ipsius dignosci, nisi sepositis corporis et spiritus rationibus; eo sibi collimandum existimavit, ut quo corpus pervenire posset, et in quo spiritibus cederet, explicatum faceret.' The Latin proverb to which he is referring is *Canis festinans caecos parit catulos*: 'The hasty dog bears blind pups.'

> are of a different power, and cannot be brought about by these instruments. You have, Andabata, a summary of that desirable book.[61]

White is building up the case for pluralism by demonstrating, from Digby, how the variety of living and non-living things arises mechanically from the mixture of elements, these latter being differentiated by the rarity or density of the *prima materia*. There is nothing in the physical world, then, except the elements and mixed bodies (White 1642, 59). In order to demonstrate that 'there is no other type of being in the universe, except those which we have numbered', White further investigates these mixed bodies following an Aristotelian method of species definition. First of all, it needs to be asked whether a particular body has only one grade of rarity or density, or whether it has many, like 'limbs'. If the latter, it is mixed; if the former, it is an element. A mixed substance, then, either grows or the quantity doesn't change. The latter is purely a mixed substance, while the former is divisible again, because either it grows only from the outside, or it has internal organs by which it achieves growth. The latter occupies the grades of plants, which can be divided again. If it only receives nutrition, then the plant is pure; but if it directs itself to nourishment and displays an active principle in its motion, it is called an animal. If the whole of the animal moves, resulting in a change of place, it is a perfect animal; if only a certain part, it should be numbered in the class of zoophytes (White 1642, 60). White concludes:

> And since you see single divisions are confined by necessity of immediate contradiction, whatever way this conjecture is resolved in the heavens, this certainly is left undoubted: no other types can be concealed under the nature of body. Therefore you see this established: this universe, that we say to be one by contiguity of size, is distributed adequately and commodiously into these of its parts.[62]

Having thus established that the observed variety of terrestrial species is in fact universal, following out of logical necessity from physical principles, the question

---

[61] White 1642, 57–58: 'Quid ergo fecit? Cum corpus idem esse quod magnum a pueris aeque ac senibus intellectum videret; inde exorsus quantitatem per rarum et densum, quasi primis differentiis scindi delineat. Ubi vero rarum sit et densum, illic adiecto, quem manibus tangimus, universi situ, ex necessitate gravitatem quoque et levitatem nasci. Ubi haec essent, primas quatuor qualitates, calorem et frigus, humorem et siccitatem abesse non posse. Et his positis, Elementorum rationem esse necessariam; quorum singula duas ex his qualitatibus Aristotelis ad iussum sortita forent. Elementa vero magna mole et suis quasi latifundiis non posse extare; particulis autem minimis continuo permisceri. Hinc mixtorum communes rationes subiungit: variorum motuum causas, et ortus exponit: ad plantas et animalia se porrigit. Ubique ostendit, ea. quae praeter hominem agantur, talia esse, ut, quasi filis ex Aranei cavo emissis et intertextis, sic e naturae visceribus et densi rarique complexionibus necessario nexu consequantur. Postea animi quoque lustrat germina, eaque alterius esse potentiae neque his instrumentis confieri posse docet. Tum demum immortalem animam, et brevem de statu separationis commentationem adiungit. Habes, Andabata, expetendi illius libelli summam…'

[62] White 1642, 60–61: 'Et quoniam immediatae contradictionis necessitate singula membra coarctari vides; qualiscumque sit coniectura haec in Coelis reperiri, hoc certe indubitatum relinquitur, alia sub corporis natura latere genera nulla posse. Hoc itaque confectum vides, universum hoc, quod contiguitate quanti unicum esse diximus, his sui partibus adaequate et commode esse dispertitum.'

turns to whether the moon specifically could sustain life.[63] Asphalius repeats Galileo's objections that the days and nights are too long, and that there is no rain. Ereunius believes that the philosophical certainty granted to him by Digby's book is sufficient reason to ignore Galileo's timid objections. If we see the solid earth of the moon surrounded by vapours, he argues, how can we deny the existence of the intermediate body, water? He refers to 'German observations' of clouds on the moon, which could well be the same claim of Michael Maestlin referred to by Kepler (see Sect. 4.5). The existence of such effluvia is later justified by White on the assumption that 'that which proves useful in the earth, is to be expected the same in those celestial globes, if, as we said, they are equal to the earth'.[64] In terms of the long days and incredible heat, White simply referred to people living at different latitudes of the earth. People in the far north endure day or night for months at a time, while those near the equator escape the heat by living at high altitudes (White 1642, 61–66). These and more explanations justify White's belief 'that indeed men and animals like ours can dwell in the moon, more happily than things are here'.[65]

So far with White we have focused on the natural philosophical basis for his pluralism, but we have not yet mentioned the other focus of our discussion: astrology. There is in fact an entire section of the dialogue devoted to the subject, but while White was happy to admit the generative effect of the sun and the moistening effect of the moon, he was highly critical of the practice of judicial astrology (White 1642, 370–82). Indeed, Mersenne, with whose circle White was associated, was motivated in his support of mechanism largely by his aversion to the astrological naturalism of Pomponazzi, Cardano and Vanini, as was mentioned above and will be discussed later with regards to Henry More (Sect. 6.2).[66] White, in his attempt to modernise Aristotelianism, would have been aware of exactly that dangerous kind of Aristotelianism promoted by the Italian naturalists, and so, being at the same time minded to rid philosophy of occult causes, he attempted to undermine the astrological rationale. Accepting the fact that he had established earlier in the dialogues, that the stars are either fires or they reflect a fiery nature received from elsewhere, he argued that any effects from celestial bodies other than the sun and the moon would be so small that they were not sensible or measurable (White 1642, 375–76). He believed that the entire origin and basis for astrology was the idolatrous belief that the celestial bodies possessed souls filled with reason and intelligence. At this point in the dialogue, the speaker Andabata, while accepting the rationality of all Ereunius' objections to astrology, has some concerns:

---

[63] This argument for extraterrestrial life from the *limited* variety of possible things is a counterpoint to Lovejoy's emphasis on the 'principle of plenitude'. It was also suggested by Knight as a reason for the surprising similarities between writers on pluralism in the seventeenth century: Knight 1967, 61–68.

[64] White 1642, 106: 'Et quod in terra usuvenit, idem in globis illis coelestibus, si telluri pares sunt, ut diximus, expectandum est.'

[65] White 1642, 62: '…non dubitem etiam homines et animalia nostratia posse in Luna, foelicius quam apud nos morari'.

[66] See Ashworth, Jr. 1986, 138.

> I see that what you have said stands on great reason. But it worries me that we read that these stars were created for the use of man, whereas if they have no effect in human things, at least not that we can tell, then I would believe this contrary to reason.[67]

It is Asphalius who replies first, giving traditional arguments for their use. They serve as signs of the seasons, months and years, and moreover as an indicator of divine magnificence, raising up souls to the contemplation of nature. They encourage a cosmic perspective, revealing the foolishness of all Alexanders and Caesars who violently divide up this speck of dust and reckon themselves the greatest of men (White 1642, 381–82). Ereunius adds to this that human utility does not exclude other ends of which we are ignorant. He counsels a safer opinion: 'This orb on which we tread we accept to be given mostly if not uniquely in our servitude; the others unknown, we leave aside, at least until a new dawn breaks from elsewhere'.[68]

A glimpse at that dawn is forthcoming, as at the end of the work Asphalius presents a proposition of truly cosmic proportions:

> The pastors also call me a prophet, so it may be allowed for me to prophesise, and if I am not able to do so concerning things far removed in time, I may at least on things distant and hidden. It seems to me that because the world is demonstrated not to be perpetual, and to be an instrument for the determination of a certain end, and this end to be something permanent and eternal—and furthermore that among bodies nothing like this exists, nor can it be framed from our elements or from quantity—it is made necessary that this whole universe was created for no other reason than on account of spirits confined in bodies, which (because we only know of one species) we call humans. All of those [celestial] bodies, therefore, which appear bright to us, if they shine of themselves, like the sun, are the least noble, and more like instruments for those [celestial bodies] which are illuminated. These latter meanwhile are the seats of the various species of human, and indeed both the quantity and measure of the universe are to be accepted thereupon: that however many possible diverse species of human there are, there are that many diverse dwelling-places constructed for them by Almighty God. And in line with the principles of the propagation of individuals, each was given a suitable structure for their house, and also proportionate periods of time…[69]

---

[67] White 1642, 381: 'Quae dixisti, video multa ratione subsistere. Sed terret me quod in usum hominum creata haec astra legimus: unde nullum in res humanas effectum, vel habere, vel certe non sensibilem, prorsus a ratione alienum crediderim.'

[68] White 1642, 382: 'Alioquin hunc quem calcamus orbem, vel maxime si non unice in nostrum famulitium deditum accipiamus; reliquos ut ignotos praetereamus, donec aliunde nova nobis albescat aurora.'

[69] White 1642, 444–46: 'Me quoque vatem pastores dicunt, liceat itaque etiam mihi vaticinari, si non valeam de longinquis tempore, certe loco distantibus et occultis, videtur mihi quod quoniam conclusum est mundum non esse perpetuum, et esse instrumentum ad determinatum finem factum, et hunc finem aliquid permanens esse, et aeternum; et inter corpora nihil tale existere; neque ex nostris elementis neque ex quantitate compingi posse; ex necessitate fieri, totum hoc universum non esse creatum, nisi propter spiritus corporibus incarceratos, quos (quia unam tantum speciem agnovimus) homines appellamus. Quot itaque nobis apparent lucida corpora, si ex sese lucent, ut sol ignobilissima esse et magis instrumentalia illis quae illuminantur, haec vero sedes esse variarum specierum hominum, atque adeo et universi quantitatem et modum inde accipiendum esse, ut quot sint hominum possibiles diversae species, tot sint illis diversa a magno numine constructa habitacula: et iuxta propagationis singulorum leges, datam molem singulis eorum tentoriis: temporis quoque proportionatos fines…'

In this passage, the philosophical principles of Digby and White reach their cosmological climax. This bold pluralism is not isolated to this one 'prophecy' at the end of *De mundo*. It was repeated again in White's *Institutiones Peripateticae* (1646), a work which was translated into English (White 1656). This later publication contained repetitions and paraphrases of much of the *De mundo*, including some more bold phrasings of the pluralist theories. There White argued that the sun must have analogous effects on the analogous matter of the other planets (White 1656, 126–27). He described the world as a 'vast wombe' in which the opaque bodies are 'cells' for the housing and nurturing of spirits (White 1656, 189–90). In his later *Contemplation of Heaven* (1654) he again talked of 'Corporeall Rationall Creatures in Globes besides the Earth' (White 1654, 93). These references cast doubt on the assertion by a modern biographer of White who, despite being familiar with some of these passages, argues that the likelihood of a plurality of worlds was diminished for theological and philosophical reasons, and that White maintained an anthropocentric stance (Southgate 1993, 102–3).

White's cosmology was certainly in one sense anthropocentric, but in the above passage from *De mundo* Asphalius makes it clear that we only call immortal souls confined in bodies 'humans' because we know only one species. White expanded on this point in his edition of the dialogues of William Rushworth (d. 1637), to which he added a preface and fourth dialogue of his own writing. In this fourth dialogue between an uncle and nephew, the nephew raises the possibility of life on other orbs in objection to the uncle's assertion that all other creatures have been made for man. 'I would not have you think I call Mankinde precisely the seed of Adam or Noe,' replies the uncle, 'but all such creatures (if there be any where others) who are due to Eternity, and yet have their births couch'd under the shels of mortality. For to all such will agree the notion of *Animal rationale*, which definition our Philosophers have determin'd to man' (Rushworth and White 1654, 224). It is apparent therefore that when White agreed that everything was created for the use of man, he was not thinking purely of terrestrial humans. Humanity does not specify shape, size, appearance or custom. Rather, in White's Aristotelian process of definition by contradiction, a human is simply a member of that branch of the Porphyrian Tree labelled 'rational animal'—in this case an immortal rational soul confined in an animal body. Digby's strict exclusion of the soul from physical considerations meant that while a universal process of abiogenesis could create ET life on the level of plants, animals and zoophytes, the creation of humans required a more direct involvement of the deity. In White's estimation, not only does the influence of the sun produce similar generations in the other planets as in the earth, but God likewise acts similarly in them, using multiple locations for the soteriological journey of humanity.

While White and Digby thus retained the importance of celestial influence for the processes of generation, White specifically precluded this from forming any basis for a wider astrological theory. The effects of the celestial bodies are general, not specific, and other than the sun and the moon they are negligible. All theories of their influence are philosophical bastardisations of idolatrous planet-worship. As we saw in Sect. 2.5 with Benedetti's letter to Pingone, the need to explain the use of the heavens was widely appreciated. White attacked the theory and practice of astrology, but knew that in doing so he would have to confront the rejoinder that God does nothing in vain, so he presented an alternate teleology of the heavens, painting the celestial bodies as other worlds filled with rational beings. The justification of this claim through Digbean philosophical principles, however, was met with incredulity by Hobbes.

5.3.3 Thomas Hobbes

Thomas Hobbes was part of the same circle in Paris as Digby, White, Mersenne, Gassendi and Descartes. The members of this philosophical community regularly responded to each other's work both privately and publicly. Shortly after the publication of White's *De mundo*, Hobbes composed a detailed refutation of most of its contents. The manuscript was only published in 1973 following its discovery in the Bibliothèque Nationale by Jean Jacquot.[70] The critique contains many ideas found in Hobbes' later *De corpore* (1655), and it has been observed that he was largely using White as a foil to sharpen his own philosophical principles (Martinich 1999, 183). This work is thus not only helpful for understanding the early development of Hobbes' mechanical philosophy, but also for demonstrating the many different possibilities under the umbrella of mechanism in general, because Hobbes seemed determined to disagree and proffer alternatives to nearly everything White said. His objections were at times physical, metaphysical or sceptical, and even occasionally theological. While his criticisms were often legitimate, his alternatives, while interesting, were hardly more convincing, as we will see.

Hobbes' very different philosophical approach is obvious in the way he tackled White's assertion that bodies analogous to those on earth exist in the heavens. He first of all contended that such an issue could not be investigated by natural reason. The extent of things which we know to exist does not in any way prevent the existence of any number of other monstrous creations, nor is there any way of telling whether the things we know do or do not exist in places unseen to us. Hobbes then drew the reader's attention to the seeming confusion of our beliefs about celestial inhabitability: we believe not only that there might be similar bodies as ours in the heavens, but also that our own bodies will one day end up there. Yet he did grasp the essential point of White's argument: whether the variety of things is necessarily finite (Hobbes 1976, 78–79). Approaching this question, Hobbes discussed Aristotle's *prima materia*, which he took to be effectively equivalent to the homogeneous atoms of modern philosophers. This *prima materia*, in his opinion, is nothing but body itself, because matter cannot be the material of itself (Hobbes 1976, 80–81). However, he believed that this material could move and change in innumerable ways, producing innumerable types of sensory images—which he took as the basis of all epistemology and hence philosophy—such that it is impossible to know how

[70] The manuscript is Bibliothèque Nationale de France, MS fonds latin 6566A.

many varieties of bodies there might be, or whether those in the heavens are like our own. It may be that they are, but then again it may be that there is nothing in the heavens even remotely comparable on a basic physical level (Hobbes 1976, 81).

Hobbes then gave a summary of White's own reasoning, based on Digbean principles and Aristotelian classification. Of this he was quite dismissive: 'Indeed, it is astonishing if, by arranging names in this way, we can deduce an argument for limiting the things we do not know and to which we have therefore not yet given names' (Hobbes 1976, 83). His last criticism on this point was that White didn't state the conclusion to which he was led, but instead left the conclusion to be inferred from the following discussion of the moon (Hobbes 1976, 84). On this subject, Hobbes agreed that telescopic observations suggest that the moon has a similar nature to the earth, and that this makes it possible that there may be living creatures there. However, he maintained that no experiment supported White's conjecture that human beings or other terrestrial animals will be found there (Hobbes 1976, 84–86). Having undermined White's philosophical limitation of bodily variety, and with no sensory experience to tell one way or another, the terrestrial nature of the moon was no basis for any conclusions about its inhabitation.

In later chapters Hobbes grew frustrated with White's teleological reasoning. When White, arguing for the existence of a central fire in the earth which contributes to generation, inferred that the earth would be useless unless man were born on it, Hobbes retorted: 'To whom, then, [one asks,] was it useful for man to exist, so that it was for his pleasure that in the centre of the earth fire was located?' (Hobbes 1976, 132). For his own part, Hobbes attributed the earth's fertility exclusively to the sun. When White, as we noted above, claimed that the other planets should have exhalations because of their parity with the earth, Hobbes contended that such equality had in no way been agreed (Hobbes 1976, 135). In a general rebuke to such life-centric teleology, he decried the belief that God could create nothing which did not benefit living things, and that human wisdom and self-esteem is any useful measure for the wisdom of God (Hobbes 1976, 136).

Hobbes' own philosophical principles were put to the test when he tackled the motions of the earth and the moon. His explanations were mechanical yet undeniably (although Hobbes did deny it) animistic; more reminiscent of Gilbert than Kepler. When attempting to explain the eccentricity of the earth's orbit, he began with the dictum that any given thing always moves towards the spot where its internal, essential motion is best conserved. It therefore maintains a distance from the sun and the other celestial bodies which is 'best fitting its own nature' (Hobbes 1976, 220). He believed that if the sun and the earth were the only two bodies, the earth would orbit in a perfect circle. Since it does not do so, this must be the effect of influences from other stars. The northern stars being more numerous than the southern, for example, could be why the earth is further from the sun when the latter body is in Cancer than when it is in Capricorn (Hobbes 1976, 221). A similar urge to even out influences underpins the earth's daily rotation. The shaded part of the earth strives to push forward in order to benefit from the sun's rays, while the illuminated side, well-sated, moves away. Hobbes explicitly likened this motion to that of living creatures when they are moved towards people or things whose behaviour aids their own inner and innate motion. 'I do not, however, think that the earth is animate', he qualified, 'yet all consistent bodies, or those possessing coherent parts, have this in common: they maintain habitual motion inasfar as other motions, gravity especially, do not hinder [them]' (Hobbes 1976, 221).

Later in this work he suggested a different reason for the earth's eccentricity: the earth, being a heterogeneous body, has some parts which need the sunlight more than others, and this leads to an inequality of motion (Hobbes 1976, 291–92). Again, this argument was supported by Hobbes with examples from animal behaviour. He anticipated the objection that such examples would only be relevant if the earth itself were an animal, a proposition he was unable to accept. Instead, he suggested a theory that was perhaps even bolder, arguing that non-living terrestrial bodies, such as logs or stones, would all rotate themselves with respect to the sun if they weren't prevented by gravity or some other force (Hobbes 1976, 292). So in fact Hobbes' theory of celestial motion is not animistic; the suggestion is rather that animate motion, and the desire which drives it, is not necessarily animate, or at least not exclusively so, being in fact inherent to all bodies. We are simply used to expressing such notions in human terms, such as appetite, pleasure, and pain, but the basis of these phenomena is universal and relates to the strengthening or obstructing of a body's specific, inner, natural motion (Hobbes 1976, 292–93).

Hobbes also applied himself to the tricky question of the motion of the moon. The upshot is that it is affected by many different bodies, such as the planets and stars, but most of all by the sun and the earth, and, being as well a heterogeneous body, its different parts are affected in different ways. The sun causes the earth and moon to orbit together once a year, while also illuminating the moon's surface. The earth drives the moon's monthly orbit by its daily rotation, but also affects it by an influence which Hobbes compared to magnetism, and again by vapours which may rise from the earth as far as the moon (Hobbes 1976, 277–78). Hobbes at one point took on this issue of influence more directly and generally. He believed that a first basic division could be made between types of influence. The influence of bodies which are homogeneous throughout, like the sun, is light. Heterogeneous bodies, however, possess as many different inner motions as there are such bodies, and these motions lead to influences distinct from light. Not only will the influence of the moon differ from that of the earth, but different parts of the earth, like metals, stones, plants, magnets, all have different influences (Hobbes 1976, 298).

The question of how exactly those influences are transmitted was dealt with later in the work when Hobbes defended the physical foundations of astrology against White's attack. From the outset, it is clear that this will not be a traditional defence of astrology. The first question he felt the need to ask was whether future events 'so depend on the present influence of the stars that they cannot fail to be brought to pass' (Hobbes 1976, 435). He used stars as a catch-all term to mean all the celestial bodies, including the earth, and the transparent air or ether between them. He next laid out the mechanical approach to the problem: all actions which are usually attributed to sympathy, antipathy, occult qualities or influences are in fact just motion (Hobbes 1976, 436). Motion is transmitted from body to body over a theoretically infinite distance, and so when a star influences the earth nothing actually departs from the star itself; the side of the star facing the earth simply presses outwards onto the air or ether, and the motion is propagated as far as the earth. This motion can be of varying kinds, and so it could potentially illuminate, heat, cool, dry or moisten. What is commonly called influence is defined by Hobbes as 'that motion in parts of the stars (the senses cannot detect it) by which a motion is propagated to a distance by a continuous thrust of the medium' (Hobbes 1976, 437).

Such an interpretation of influence was much more amenable to the doctrine of aspects, which Hobbes duly defended on the basis that contrary motions retard each other, while conjoined motions strengthen one another. The other aspects will vary along this continuum. Whether the earth is the centre of the planetary orbits or not makes no difference to either the virtues and aspects of the celestial bodies, and we need not worry about the recent discovery of the satellites of Jupiter and Saturn: 'Undoubtedly, even before they became visible to man, those secondary planets influenced Jupiter and Jupiter them' (Hobbes 1976, 438). Indeed, influences are not just coming from the stars to the earth, but are also being exchanged between stars, while the earth itself exerts an influence on distant bodies. Everything is continually acting on everything else, and it is the combination of all these influences at any one time which produces a particular effect. This is the determinism for which Hobbes is so well-known: 'Therefore any effect within any body must be wholly ascribed to the virtue of the stars. Hence the sum total of all the causes within the stars is the sum total of all the causes within the universe' (Hobbes 1976, 437).

Hobbes argued that White himself conceded this point on necessity and astral causality, and indeed White did seem to do so, not only in the section on Stoicism in the *De mundo*, but also in his later *Contemplation* (White 1642, 359–61, 1654, 77–82). Yet White's idea of fate involved God foreseeing the free decisions of every future person and planning accordingly. Hobbes, on the other hand, echoing the naturalism of Pomponazzi and Vanini, refused to exempt human decisions from the chain of secondary natural causes. He dealt with this directly when countering White's argument that astrology relies on a belief in animate celestial bodies:

> God performs all natural actions, the voluntary no less than the involuntary, by means of second causes, namely the bodies constituting the universe. Therefore all actions should equally pre-exist in second causes, and hence all the appetites of animals, i.e. the wishes of men, should also possess their own causes composed of the gathering together of everything requisite to volition. That is, appetites should necessarily originate in secondary causes, be these animate or not-animate. So, to [produce] voluntary actions or those ordered by a kind of purpose leading towards a defined end, it is unnecessary for every agent to be animate, much less [for it] to possess intelligence. Many species of animals originate from putrefying matter, yet their actions are marked out and ordered by design towards a specific end; but neither the sun (that generates them) nor the decaying material is on that account to be considered as possessing intellect (Hobbes 1976, 439–40).

Hobbes refused to accept that the processes which lead to human action are ontologically distinct from those of animals, which, as everyone knows, can arise from unintelligent matter. Where White argued that free will is what makes mankind worthy to be the cause for which God created the universe, Hobbes replied that such merit is shared with the animals. He understood that people are worried about admitting an external cause for the will, because it could then be invoked to cover all sins. People who thought otherwise, like Hobbes himself, upheld the view of an external cause not only because 'for the celestial and eternal chain of causation to be broken is an affront to the Divine Majesty; but also that [for this to happen] is against natural reason' (Hobbes 1976, 458).

In the end Hobbes declared that the arguments used by White against astrology were invalid, but then, after confirming the inescapability of celestial causation, he declared that the infinitude and imperceptibility of stellar motions made comprehension, and hence prediction, impossible (Hobbes 1976, 440). He did suggest that weather prediction might be possible if sufficient records and commentaries were kept, but these 'neither exist nor will exist' (Hobbes 1976, 441). So why did Hobbes go to the effort of dismantling White's arguments against judicial astrology, only to dismiss it himself in the end? One answer is of course the contrarian nature of the work. Another is that it gave Hobbes a chance to present his arguments for mechanical determinism and the material nature of volition. Yet throughout his discussion of celestial motions and influence there is an emphasis on causes, influences and effects being shared and universal, in a way which had been played down when he criticised White's arguments for ET life. The surety of his theorising didn't quite square with his earlier scepticism.

One clue to working this out lies in Hobbes' ideas about epistemology and the value of philosophy. Consider this passage from his *De corpore*:

> The End or Scope of Philosophy, is, that we may make use to our benefit of effects formerly seen; or that by application of Bodies to one another, we may produce the like effects of those we conceive in our minde, as far forth as matter, strength and industry will permit, for the commodity of humane life. For [t]he inward glory and triumph of mind that a man may have, for the mastering of some difficult and doutfull matter, or for the discovery of some hidden truth, is not worth so much paines as the study of Philosophy requires; nor need any man care much to teach another what he knowes himselfe, if he think that will be the onely benefit of his labour. The end of Knowledge is Power; and the use of Theoremes (which among Geometricians serve for the finding out of Properties) is for the construction of Problemes; and lastly, the scope of all speculation is the performing of some action, or thing to be done (Hobbes 1656, 5).

In the same work, astrology is excluded from philosophy not because it is false, but because it is not well grounded (Hobbes 1656, 8). Hobbes believed that the reality of celestial influence could be proved by demonstrable philosophical and geometrical principles. While we can be sure that nothing happens that is not caused 'by the several motions of all the several things that are in the World', that brings us no closer to predicting future events (Hobbes 1656, 393). Similarly, while we may concede the possibility, for example, of some sort of living creatures on the moon, without direct sensory experience we have no rational means of making conjectures. Nor, in Hobbes' opinion, would such knowledge be of any use. Perhaps that is why he employed a definite mocking tone when he considered Asphalius' final statement in *De mundo* that there are as many races of men in the universe as there are externally illuminated celestial bodies. 'Such questions, however, are too slight to be argued against', he wrote, 'and whether they are true we shall not be in a position to say before the human race (which has happened to secure its dwelling on earth) has found some way of conversing with the rest of the species of living beings: the inhabitants of the moon and of the other planets' (Hobbes 1976, 497).

5.4 Conclusion

This chapter began with a survey of influence and inhabitation being employed together, and then presented a snapshot of one corner of early mechanism to understand how each of these ideas might come under debate. This transition involved a step back in time, and in the writings of Digby, White and Hobbes we came across certain ideas and themes related to astrology and pluralism which are familiar from the work of other astronomers and natural philosophers both before and after them. Digby was another, like Patrizi and Gilbert, who equated light with fire, but his description of its role in generation, both spontaneous and otherwise, took on a new atomist bent. He also stated more definitively that the action of light on the other planets should produce similar results. This belief in the analogy between terrestrial and planetary material and the universal action of light was gaining wider credence. The astronomer Ismaël Boulliau (1605–1694) considered this to be the main purpose of sunlight in opposition to Kepler's theory that it was responsible for the motion of the planets:

> …the rays and active virtue of the sun act in the planets through heat and light not so as to move, but so as to produce alterations, and, if such is their nature, generations and corruptions to some extent, such as it produces in the earth.[71]

As for an actual atomist account of such generations, there were as many theories as there were philosophers. The question of teleology is, of course, as entrenched in biology as it is in astronomy, if not more so.[72] Digby, as we have seen, described seeds as the product of a natural process of distillation, while allowing for spontaneous generation without the presence of a seed. Another early English proponent of atomism, Walter Charleton (1619–1707), preferred the Epicurean theories of Gassendi, which described small molecules of atoms, variously composed and configured, as the seminaries of various productions (Charleton 1654, 105).[73] These two ideas would be continually debated in the following century, especially in regard to the pre-existing germ theory.[74]

Digby also gave a role in generation to the 'universal spirit', a more general distillation of life-giving virtues produced by the reflection of the sun's rays against all

[71] Boulliau 1645, 22: 'Tertiam vero nihil aliud probare dico, nisi Solis radios, et virtutem activam, per calorem, et lucem in planetas agere, non ut moveat, sed alterationes, et si talis sit eorum natura, generationes, et corruptiones quadamtenus operetur, quales ipse in terra producit.'

[72] See Osler 2001.

[73] On Gassendi's theory of homogenesis, or pre-disposed matter, see Fisher 2003. For a more general view, see Chang 2011.

[74] See Farley 1977; Smith 2006.

bodies within its reach. In this sense, his philosophy is an example of the symbiosis of astrological and pluralist thinking explored in the first half of this chapter. Charleton agreed at least with the principle that light could take on the attributes of any opaque or diaphanous material it encountered, thus giving it new and particular influences over the process of generation.[75] Yet Charleton was of a similar opinion to White that astrology usurped divine providence and human free will (Charleton 1654, 349). Hobbes, as we have seen, thought quite differently. In terms of astronomical cosmology, it was Digby and White who were more in line with the development of modern thought. Regarding pluralism, Hobbes' warning of the limits of observation and possible conjecture were largely ignored by future philosophers who embraced the idea of a multiply inhabited universe, the basic homogeneity of which was guaranteed by the limits of physical possibility. His defence of a naturalist astrological determinism, meanwhile, saw him surpass the likes of Vanini, Cardano and Pomponazzi as the philosophical bête noire and atheist *par excellence* of the seventeenth century.

The question of whether astrology confirms or denies God's providence is almost as old as astrology itself, and we will visit it again soon. What became more prevalent in the mid- to late-seventeenth century, as our earlier survey demonstrated, was an idea on which both White and Hobbes agreed: that perhaps not everything in the universe was created for the use of the inhabitants of the earth. While Hobbes would go so far as to suggest that God could create objects without regard to any living thing, most, like White, felt the urge to preserve biological utility, and pluralism was the obvious substitute. In Gassendi's own attack on astrology, he countered the charge of futility by suggesting that the celestial bodies 'serve among themselves to that further end, which the infinite wisdom intended at their Creation' (Gassendi 1659, 125). Gassendi, like Descartes and Galileo, was critical of overly-anthropocentric cosmological thinking of the kind implicit in astrology, but was reluctant to endorse theories about ET life. Others were not so restrained. In the late seventeenth and early eighteenth centuries we see the symbiosis of astrology and pluralism deteriorate, as astrology continues to come under attack—and, perhaps more significantly, is avoided or ignored—while pluralism becomes ingrained in the teleology of a Newtonian cosmos.

References

Ashworth, William B., Jr. 1986. Catholicism and early modern science. In *God and nature: Historical essays on the encounter between Christianity and science*, ed. David C. Lindberg and Ronald L. Numbers, 136–166. Berkeley: University of California Press.

Bachrach, A.G.H. 1987. Luna mendax: Some reflections on moon-voyages in early seventeenth-century England. In *Between dream and nature: Essays on utopia and dystopia*, ed. Dominic Baker-Smith and C.C. Barfoot, 70–90. Amsterdam: Rodopi.

Borel, Pierre. 1657. *Discours nouveau prouvant la pluralité des mondes*. Geneva: s.n.

[75] See his discussion of the qualities of sunlight reflected from the moon: Charleton 1654, 352.

———. 1658. *A new treatise, proving a multiplicity of worlds*. Trans. D. Sashott. London: John Streater.

Boulliau, Ismaël. 1645. *Astronomia philolaica*. Paris: Simeon Piget.

Bowden, Mary Ellen. 1974. The scientific revolution in astrology: The English reformers, 1558–1686. PhD dissertation, Yale University.

Boyle, Robert. 1999. *The works of Robert Boyle*, ed. Michael Hunter and Edward B. Davis, 14 vols. London: Pickering and Chatto.

Campbell, Mary B. 1999. *Wonder & science: Imagining worlds in early modern Europe*. Ithaca/London: Cornell University Press.

Carré, Marie-Rose. 1974. A man between two worlds: Pierre Borel and his *Discours nouveau prouvant la pluralité des mondes* of 1657. *Isis* 65: 322–335.

Chang, Ku-ming (Kevin). 2011. Alchemy as studies of life and matter: Reconsidering the place of vitalism in early modern chymistry. Isis 102: 322–329.

Charleton, Walter. 1654. *Physiologia Epicuro-Gassendo-Charltoniana, or, a fabrick of science natural, upon the hypothesis of atoms founded by Epicurus*. London: Thomas Heath.

Clericuzio, Antonio. 2000. *Elements, principles, and corpuscles: A study of atomism and chemistry in the seventeenth century*. Dordrecht/London: Kluwer Academic Publishers.

Clucas, Stephen. 1995. *Thomas Harriot and the field of knowledge in the English renaissance*. Oxford: Oriel College.

———. 1997. "The infinite variety of formes and magnitudes": 16th- and 17th-century English corpuscular philosophy and Aristotelian theories of matter and form. *Early Science and Medicine* 2: 251–271.

———. 2001. Corpuscular matter theory in the Northumberland circle. In *Late medieval and early modern corpuscular matter theories*, ed. Christoph Herbert Lüthy, John Emery Murdoch, and William Royall Newman, 181–207. Leiden: Brill.

Cressy, David. 2006. Early modern space travel and the English man in the moon. *The American Historical Review* 111: 961–982.

Davidson, Nicholas S. 2005. "Le plus beau et le plus meschant esprit que ie aye cogneu": Science and religion in the writings of Giulio Cesare Vanini, 1585–1619. In *Heterodoxy in early modern science and religion*, ed. John Hedley Brooke and Ian Maclean, 59–80. Oxford/New York: Oxford University Press.

Dick, Steven J. 1982. *Plurality of worlds: The origins of the extraterrestrial life debate from Democritus to Kant*. Cambridge/New York: Cambridge University Press.

Digby, Kenelm. 1645 [1644]. Two treatises: In the one of which, the nature of bodies; in the other, the nature of mans soule, is looked into: In way of discovery of the immortality of reasonable soules ... London: John Williams.

———. 1658. *A late discourse touching the cure of wounds by the powder of sympathy*. Trans. Robert White. London: R. Lownes and T. Davies.

———. 1661. *A discourse concerning the vegetation of plants*. London: John Dakins.

Dobbs, Betty Jo. 1971. Studies in the natural philosophy of Sir Kenelm Digby. *Ambix* 18: 1–25.

———. 1973. Studies in the natural philosophy of Sir Kenelm Digby part II. Digby and alchemy. *Ambix* 20: 143–163.

———. 1974. Studies in the natural philosophy of Sir Kenelm Digby part III. Digby's experimental alchemy – The book of secrets. *Ambix* 21: 1–28.

Ernst, Germana. 2015. Maculae Galilei me perplexum habent.' Campanella, Sun-spots and Pythagorean temptations. In *Authority, innovation and early modern epistemology: Essays in honour of Hilary Gatti*, ed. Martin McLaughlin, Ingrid D. Rowland, and Elisabetta Tarantino, 170–185. Cambridge: Legenda.

Farley, John. 1977. *The spontaneous generation controversy from Descartes to Oparin*. Baltimore: Johns Hopkins University Press.

Fisher, Saul. 2003. Gassendi's atomist account of generation and heredity in plants and animals. *Perspectives on Science* 11: 484–512.

Gadroys, Claude. 1674. *Discours physique sur les influences des astres selon les principes de M. Descartes*. Paris: Jean Coignard.
Galilei, Galileo. 1890–1909. *Opere*, ed. Antonio Favaro, 20 vols. Florence: G. Barbèra.
Gassendi, Pierre. 1659. *The vanity of judiciary astrology. Or divination by the stars. Lately written in Latine, by that great schollar and mathematician the illustrious Petrus Gassendus; mathematical professor to the king of France*. Translated into English by a person of quality. London: Giles Calvert.
Gilbert, William. 1651. *De mundo nostro sublunari philosophia nova*. Amsterdam: Lodewijk Elzevir.
Godwin, Francis. 1638. *The man in the moone*. London: John Norton.
Grant, Edward. 1974. *A source book in medieval science*. Cambridge, MA: Harvard University Press.
Havers, G., trans. 1664. *A general collection of discourses of the virtuosi of France*. London: Thomas Dring and John Starkey.
Henry, John. 1982. Atomism and eschatology: Catholicism and natural philosophy in the Interregnum. *The British Journal for the History of Science* 15: 211–239.
———. 1986. Occult qualities and the experimental philosophy: Active principles in pre-Newtonian matter theory. *History of Science* 24: 335–381.
———. 2008. The fragmentation of renaissance occultism and the decline of magic. *History of Science* 46: 1–48.
———. 2010. Sir Kenelm Digby, recusant philosopher. In *Insiders and outsiders in seventeenth-century philosophy*, ed. G.A.J. Rogers, Tom Sorell, and Jill Kraye, 43–75. New York/London: Routledge.
Hill, Nicholas. 1619 [1601]. *Philosophia Epicurea, Democritiana, Theophrastica proposita simpliciter, non edocta*, 2nd ed. Geneva: François Le Fèvre.
Hobbes, Thomas. 1656. *Elements of philosophy, the first section, concerning body*. Anonymous translation. London: R. and W. Leybourn.
———. 1973. *Critique du* De mundo *de Thomas White*, ed. Jean Jacquot and Harold Whitmore Jones. Paris: J. Vrin.
———. 1976. *Thomas White's* De mundo *examined*. Trans. Harold Whitmore Jones. London: Bradford University Press.
Hutchison, Keith. 1982. What happened to occult qualities in the scientific revolution? *Isis* 73: 233–253.
Hutton, Sarah, ed. 1990. *Henry More (1614–1687) tercentenary studies*. Dordrecht: Kluwer Academic Publishers.
———. 2005. The man in the moone and the new astronomy: Godwin, Gilbert, Kepler. *Etudes epistémè* (7): 3–13.
———, ed. 2017. Cambridge Platonism. Special issue. *British Journal for the History of Philosophy* 25.5.
Jacquot, Jean. 1974. Harriot, Hill, Warner and the new philosophy. In *Thomas Harriot; Renaissance scientist*, ed. John William Shirley, 107–128. Oxford: Clarendon Press.
Jaki, Stanley L. 1978. *Planets and planetarians: A history of theories of the origin of planetary systems*. Edinburgh: Scottish Academic Press.
Kargon, Robert Hugh. 1966. *Atomism in England from Hariot to Newton*. Oxford: Clarendon Press.
Knight, David. 1967. Uniformity and diversity of nature in seventeenth century treatises on plurality of worlds. *Organon* 4: 61–68.
Knobloch, Eberhard. 2003. Otto von Guericke und die Kosmologie im 17 Jahrhundert. *Berichte zur Wissenschaftsgeschichte* 26: 237–250.
Koyré, Alexandre. 1957. *From the closed world to the infinite universe*. Baltimore: Johns Hopkins Press.

Le Grand, Antoine. 1694 [1671. *An entire body of philosophy, according to the principles of the famous Renate Des Cartes, in three books. Translated by Richard Blome*. London: Samuel Roycroft.

Lobis, Seth. 2011. Sir Kenelm Digby and the power of sympathy. *Huntington Library Quarterly* 74: 243–260.

Lovejoy, Arthur O. 1948 [1936]. The great chain of being: A study of the history of an idea. Cambridge, MA: Harvard University Press.

Maclean, Ian. 2005. Heterodoxy in natural philosophy and medicine: Pietro Pomponazzi, Guglielmo Gratarolo, Girolamo Cardano. In *Heterodoxy in early modern science and religion*, ed. John Hedley Brooke and Ian Maclean, 1–30. Oxford/New York: Oxford University Press.

Martinich, A.P. 1999. *Hobbes: A biography*. Cambridge: Cambridge University Press.

Massa, Daniel. 1977. Giordano Bruno's ideas in seventeenth-century England. *Journal of the History of Ideas* 38: 227–242.

McColley, Grant. 1939. Nicholas Hill and the *Philosophia epicurea*. *Annals of Science* 4: 390–405.

Millen, Ron. 1985. The manifestation of occult qualities in the scientific revolution. In *Religion, science, and worldview: Essays in honor of Richard S. Westfall*, ed. Margaret J. Osler and Paul Lawrence Farber, 185–216. Cambridge/New York: Cambridge University Press.

More, Henry. 1646. *Democritus Platonissans: Or, an essay upon the infinity of worlds out of Platonick principles*. Cambridge: printed by Roger Daniel.

Nicolson, Marjorie Hope. 1948. *Voyages to the moon*. New York: Macmillan.

Osler, Margaret J. 2001. Whose ends? Teleology in early modern natural philosophy. *Osiris* 16: 151–168.

Poole, William. 2005. The origins of Francis Godwin's *The man in the moone* (1638). *Philological Quarterly* 84: 189–210.

Principe, Lawrence M. 2013. Sir Kenelm Digby and his alchemical circle in 1650s Paris: Newly discovered manuscripts. *Ambix* 60: 3–24.

Renaudot, Théophraste. 1655. *Recueil general des questions traittées és conferences du Bureau d'adresse és années 1633. 34. 35. jusques à present, sur toutes sortes de matieres, par les plus beaux esprits de ce temps. Tome Second*. Paris: Guillaume Loyson.

Roberts, Adam. 2006. *The history of science fiction*. Basingstoke: Palgrave Macmillan.

Roger, Jacques. 1986. The mechanistic conception of life. In *God and nature: Historical essays on the encounter between Christianity and science*, ed. David C. Lindberg and Ronald L. Numbers, 277–295. Berkeley: University of California Press.

Romanowski, Sylvie. 1998. Cyrano de Bergerac's epistemological bodies: "Pregnant with a thousand definitions." *Science Fiction Studies* 25: 414–432.

Roos, Anna Marie Eleanor. 2001. *Luminaries in the natural world: The sun and the moon in England, 1400–1720*. New York/Oxford: Peter Lang.

Rushworth, William, and Thomas White. 1654. *Rushworth's dialogues, or the judgement of common sence in the choice of religion. Corrected and inlarged by Thomas White Gentl.* Paris: Jean Billaine.

Rutkin, H. Darrel. 2006. Astrology. In *The Cambridge history of science. Vol. 3, Early modern science*, ed. Katharine Park and Lorraine Daston, 541–561. Cambridge: Cambridge University Press.

Schimank, Hans. 1967. Traits of ancient natural philosophy in Otto von Guericke's world outlook. *Organon* 4: 27–37.

Sherburne, Edward. 1675. *The sphere of Marcus Manilius made an English poem: With annotations and an astronomical appendix*. London: Nathaniel Brooke.

Smith, Justin E.H., ed. 2006. *The problem of animal generation in early modern philosophy*. Cambridge: Cambridge University Press.

Southgate, Beverley C. 1993. *"Covetous of truth": The life and work of Thomas White, 1593–1676*; Dordrecht/Boston: Kluwer Academic Publishers.

Spink, Aaron. 2017. *Cartesian method and experiment*. PhD dissertation, University of South Florida.

Thomas, Keith. 1971. *Religion and the decline of magic: Studies in popular beliefs in sixteenth and seventeenth century England.* London: Weidenfeld & Nicolson.

Thorndike, Lynn. 1923–1958. *A history of magic and experimental science*, 8 vols. New York: Columbia University Press.

Trevor-Roper, H.R. 1989. Nicholas Hill the English atomist. In *Catholics, Anglicans and puritans: Seventeenth-century essays*, ed. H.R. Trevor-Roper, 1–39. London: Fontana Press.

van Lansberge, Philippe. 1630. *Commentationes in motum terrae diurnum, & annuum.* Trans. Martinus Hortensius. Middleburg: Zacharias Roman.

Vanini, Lucilio. 1616. *De admirandis naturae reginae deaeque mortalium arcanis libri quatuor.* Paris: Adrien Perier.

Vermij, Rienk. 2002. *The Calvinist Copernicans: The reception of the new astronomy in the Dutch Republic, 1575–1750. Chicago.* London: University of Chicago Press.

———. 2016. Seventeenth-century Dutch natural philosophers on celestial influence. In *Unifying heaven and earth: Essays in the history of early modern cosmology*, ed. Miguel Á. Granada et al., 291–315. Barcelona: Edicions Universitat Barcelona.

von Guericke, Otto. 1672. *Experimenta nova (ut vocantur) Magdeburgica de vacuo spatio.* Amsterdam: Johann Jansson.

———. 1994. *The new (so-called) Magdeburg experiments of Otto von Guericke.* Trans. Margaret Glover Foley Ames. Dordrecht/Boston: Kluwer Academic.

White, Thomas. 1642. *De mundo dialogi tres.* Paris: Moreau.

———. 1654. A contemplation of heaven with an exercise of love, *and a descant on the prayer in the garden.* Paris: s.n.

———. 1656. *Peripateticall institutions. In the way of that eminent person and excellent philosopher Sr. Kenelm Digby. The theoricall part. Also a theologicall appendix of the beginning of the world.* London: R. D.

Wilkins, John. 1638. *The discovery of a world in the moone.* London: Michael Sparke and Edward Forrest.

Wittie, Robert. 1681. *Ouranoskopia. Or, a survey of the heavens: A plain description of the admirable fabrick and motions of the heavenly bodies, as they are discovered to the eye by the telescope, and several eminent consequences illustrated thereby.* London: J.M.

Chapter 6
Influence and Inhabitation Opposed

On the other Hand, how is it possible to conceive that, that immense Number of glorious and Sun-like Bodies of the fixt Stars, those vast and huge Bodies of some of the Planets (in respect of our Earth) with their noble Attendance, were made for no other use but to twinkle to us in Winter Evenings, and by their Aspects to forebode what little Changes of Weather, or other pitiful Accidents were to be expected below, or to be peep'd at by some poor Paltry Fellows of Astronomers?

(Cheyne 1705, 110)

Abstract This chapter begins with a brief description of some anti-pluralist writings from the seventeenth century. The argument is made that while a belief in pluralism rarely meant abandoning astrological thinking, the outright denial of pluralism did necessarily involve the reaffirmation of an astrological teleology. The first main section of the chapter analyses a debate between Henry More and John Butler, an Anglican minister, about the legitimacy of astrology. The chapter then continues with a discussion of how pluralism was starting to take over teleological ground from astrology. The last section of the chapter concentrates on the program of natural theology, arguing that there was a conscious effort to deny the role of celestial influence in generation, and to re-orient the understanding of God's providence around a widely populated cosmos rather than an astrological one.

Keywords Astrology · Extraterrestrial life · Henry More · Teleology · John Flamsteed · Bernard Fontenelle · Christiaan Huygens · Natural theology · Richard Bentley · Newtonian

© Springer Nature Switzerland AG 2019

J. E. Christie, *From Influence to Inhabitation*, International Archives of the History of Ideas Archives internationales d'histoire des idées 228,
https://doi.org/10.1007/978-3-030-22169-0_6

6.1 Introduction

While Hobbes' criticism of pluralism centred on its lack of utility and philosophical exactitude, there were others in the seventeenth century whose opposition to the rising popularity of ET life was more direct and explicit. The English clergyman Alexander Ross (1591–1654) was one who wrote several works against the new astronomy. His *Commentum de terrae motu circulari* ('The Fiction of the Circular Motion of the Earth', 1634) was directed largely against Lansberge's similarly titled work, while his *The New Planet No Planet* (1646) was a response to Wilkins' lunar work. In response to the pluralist claims of Wilkins' book, Ross reaffirmed an anthropocentric astrological worldview. One of his first attacks on the extravagant conceits of the new philosophy involved the claim that we must find 'some office' for the earth if it is to be considered a planet (Ross 1646, fol. A3^{r}). The ridiculousness of such a proposition he considered self-evident. The earth, for Ross, was purely a receiver and not an emitter of celestial influence.

This was perfectly in line with Ross' anti-Copernican agenda. The earth was placed in the middle of the world not only because it is the most honourable place and thus befitting man as lord of the universe, but also because the centre is the best place for it to receive 'equall comfort and influence' from all parts of the heavens (Ross 1646, 58). Just like a seed grows in the centre of the fruit where 'all the powers of the plant meet together', so do all the powers of the universe unite together in the earth 'as in a small epitome' (Ross 1646, 58). Ross constantly reasserted the dogma that everything in the universe, including the sun, planets and even the new stars discerned by the telescope, was intended solely for man's use. On these grounds, he rejected new astronomical opinions about the motion of the earth and the size of the space between Saturn and the fixed stars (Ross 1646, 67). Such idle and vain speculations distract from the good uses of astronomy in medicine, agriculture, time-keeping, the prediction of eclipses, and 'such things as have their immediate dependence from the opposition and conjunction of starres' (Ross 1646, 67).

We have already mentioned the Jesuit Athanasius Kircher, and we will do so again before long. His *Itinerarium exstaticum* (1656), depicting a journey through the heavens guided by an angel called Cosmiel, was a widely read cosmological work. It showed the planets as elemental bodies complete with geological formations, and even had the stars as suns with their own orbiting planets, but not one of these celestial bodies had any human, animal or vegetable life; such a privilege was reserved for the earth.[1] Instead, Kircher portrayed the surfaces of the planets as embodiments of their astrological qualities, and hence that description served as a physical explanation for said qualities. While such reservation about ET life seems conservative in the context of this discussion, Kircher was in fact quite progressive, or perhaps we should say eccentric, for a Jesuit writing in Rome. Although the intellectual restriction of the Counter-Reformation is often overstated, it is true that

[1] Later editions of this work had altered titles. See Kircher 1660, 45, 341. See also Fletcher 1970; Glomski 2015.

works promoting pluralist ideas were censored, prohibited and retroactively condemned.[2]

In France one of the loudest voices against Copernicanism and pluralism was that of Jean-Baptiste Morin (1583–1656). Morin was professor of mathematics at the Collège Royal in Paris from 1630 until his death, and from there he directed attacks against Galileo and Descartes and worked on the 26 books of his *Astrologia Gallica*, the 850 pages of which were only published posthumously in 1661. While the practical aspects of astrology outlined in this work were largely orthodox, there were several concessions to recent philosophical developments which were un-Aristotelian and non-traditional. He stated, for example, that the planets are mixed bodies. The question then remained whether they are mixtures of ethereal or elemental material. Recent observations suggested that the first option is out, but the alternative was not acceptable either:

> For the planets are the universal causes of all mixtures that are elemental and of man himself, by the common consensus of philosophers, hence the saying of Aristotle: 'Man is begotten by man and the sun'. Therefore they are not of the same type as those mixed bodies, because for something to be its own cause is absurd.[3]

The truth must therefore be that the planets are mixtures of both celestial and elemental material, and in each of them a different one of the elements predominates, according to and accounting for their diverse natures (Morin 1661, 171).

In his chapter on the particular mixture of the moon he first recounted the opinions of Anaximenes, Parmenides, Pythagoras, Heraclitus, Plato, Democritus, Xenophanes, Pythagoreans, and then Plutarch. The last, he knew, was used by Copernicans to argue for both the motion of the earth and the inhabitation of the moon (Morin 1661, 173). The proponent of lunar inhabitation singled out by Morin was Kepler, and the astrologer dedicated several dense pages to a long extract from Kepler's *Somnium* followed by an equally long refutation (Morin 1661, 174–78). The final conclusion was definitive:

> It is therefore also alien from reason and truth for the other globes of the world to be habitats for rational creatures, much less irrational creatures. Indeed, all the celestial bodies were created for this end only (besides recounting the glory of God): by their light, heat, cold, influence and motion to serve man, for the grace of whom alone the whole corporeal world was made before man himself; and after all man himself is the end of this world. Thus we see how many ingrates deny themselves God's great favour.[4]

---

[2] Apart from the obvious cases of Copernicus, Bruno and Kepler, other works containing heterodox cosmological views were condemned, such as Palingenius' *Zodiacus vitae* (1536) and Patrizi's *Nova de universis philosophia* (1591).

[3] Morin 1661, 170: 'Nam Planetae causae sunt universales mixtorum omnium duntaxat Elementalium, ipsiusque hominis de communi Philosophorum consensu, unde illud Aristotelis, Sol et homo generant hominem, proindeque non sunt de genere mixtorum ipsorum; alioquin essent etiam sui causa, quod absurdum est.'

[4] Morin 1661, 177: 'Ergo etiam a ratione et veritate alienum est, alios Mundi globos creaturis rationalibus, et multo minus solis irrationalibus esse habitatos. Sed corpora omnia Coelestia eo tantum fine facta sunt, ut praeter Dei gloriam quam enarrant, suo lumine, calore, frigore, influxu, atque motu homini deserviant, cuius solius gratia, conditus est totus Mundus corporeus ante hominem

These few examples suggest that while a belief in pluralism did not always mean abandoning astrological thinking (although it could), the outright denial of pluralism did necessarily involve a reaffirmation of astrological teleology.[5] The impossibility of a superfluous creative act of God, the reticence of most cosmologists to leave questions unanswered, and the strong belief in the centrality of living organisms to God's creative purpose meant that astronomy and biology would meet in either astrology, ET life, or both.

6.2 The More/Butler Debate

The paradigms of influence and inhabitation were placed in direct opposition within a dispute over astrology between Henry More and the Anglican minister John Butler (1626–1698).[6] We have already discussed the pluralist cosmology of More's *Democritus Platonissans* (see Sect. 5.2). Since the publication of that work he had distanced himself from Descartes and argued for a strict dualism which identified matter as inert, and gave the credit for all activity in the universe to an immaterial 'spirit of nature'. This position was outlined in *The Immortality of the Soul* (1659). His criticism of astrology, meanwhile, was contained in four chapters of *An Explanation of the Grand Mystery of Godliness* (1660). There he singled out Pomponazzi, Cardano and Vanini for demonstrating the atheistic and blasphemous potential of this dangerous art. The worst crimes of these astrologers were casting a horoscope for Christ and explaining the rise and fall of religions by the changing configurations of the heavens. The idea that our salvation may be subject to celestial influence or indeed any natural law was abhorrent to More (More 1681, vi).

Butler, by his own account, first encountered these chapters in 1671, while his *Christologia*, in which he himself used astrology to calculate Christ's date of birth, was in the press. That work already contained a defence of astrology in a postscript directed against John Selden (1584–1654), who had also published a work on Christ's birthdate 10 years previous.[7] Butler was urged to pull his work from the

---

ipsum; et post omnia homo ipse, huius Mundi finis; utcunque non pauci ingratissimi tantum erga se Dei beneficium negent.'

[5] It has recently been argued that the use of astrology as a basis for arguments against the new astronomy served only to maroon it in the old cosmology. Figures such as Morin thus precipitated the downfall of astrology by provoking leading figures of the new philosophy, such as Pierre Gassendi, to attack its theoretical foundations. See Hatch 2017.

[6] Those dates are based on the assumption that this John Butler, B.D., author of *Christologia* (London, 1671) and *Hagiastrologia* (London, 1680), made rector of Litchborough in 1651, is the same as the John Butler, B. D., who was married to Martha Perkins in nearby Weedon Bec in that same year, and later wrote *The True State of the Case* (London, 1697) to defend himself from claims of adultery. Other biographical details support this identification. See Butler 1697; Baker 1822–1841, I, 409–10; Burke 1835–1838, III, 253; Foster 1891, 222.

[7] Selden 1661. Anthony Wood had noted this postscript in his entry on Selden in the *Athenae Oxonienses*, calling Butler a pretender to the art of astrology and assuming that he was a Cambridge

press due to its blasphemous implications and More's anti-astrological chapters were presented to him as evidence (Butler 1680, 5). Instead of retracting his work, Butler doubled down and wrote another vindication of astrology in response to More's attacks which was published in 1680. More himself considered the matter serious enough to then reprint the four chapters of his *Grand Mystery* the following year as *Tetractys Anti-Astrologica* (1681) with notes summarising and responding to Butler's defence.

It is worth mentioning here that the period in England from the 1640s to the 1660s has been described as the 'halcyon days' of astrology (Curry 1989, 19–22).[8] During the political upheavals of the Civil War astrological predictions abounded and the ensuing break-down of royal censorship saw a huge growth in astrological publications. Professional astrologers, the most notorious being William Lilly (1602–1681), gained wealth and renown, and thus attracted the attention and scorn of many religious and intellectual figures.[9] This new class of largely self-taught practitioners were branded as soothsayers and charlatans, and the connection to politics in particular was seen as a dangerous threat. It is in this context that we should view the attacks of White and More, and indeed the defence of Butler, which are only a few examples of a wide-ranging and often bitter debate over the moral, religious and philosophical legitimacy of astrology against the background of its burgeoning popularity and social impact. We have noted already Hobbes' attack on the predictive arts in his political philosophy, even though his deterministic physics supported at least the principle of celestial causation. For More, on the other hand, what worried him most was the atheistic and irreligious implications of astrology on both a practical and theoretical level.

The combatting of atheism in physics and metaphysics was one of, if not the main concern of More's philosophical career.[10] In the *Tetractys* he outlined two main varieties of atheism: Epicurean and Aristotelian (More 1681, 3). The former was what led to his distancing himself in his later career from the mechanical philosophy because of the tendency of some, especially Hobbes, to deny the existence of any incorporeal substance.[11] The latter brand of atheism, while allowing spiritual substance, was guilty of not allowing God any direct influence on the world, assigning everything, including miracles and prophecies, to the secondary causation of the celestial spheres. Even if one such as Vanini posited a single immaterial 'soul of the heavens', it was still limited to act via the regular and predictable influences of the stars and planets (More 1681, 4). More himself believed that these pagan supersti-

man. Butler actually graduated B. D. from Trinity College Oxford in 1660. See Wood 1691–1692, I, 110; Foster 1891, 222.

[8] See also Thomas 1971, chaps 10–12; Monod 2013, chap. 2.

[9] See Geneva 1995.

[10] For analysis of the anti-atheistic mission of More and other Cambridge Platonists, especially in terms of their refutations of Hobbes, see the chapter 'Anti-Atheist Plato' in Sheppard 2015, 137–81. For an introduction to atheism and the new science, and on More's dispute with Boyle, see Henry 1990; Henry 2009. On Henry More and the nature of spiritual extension, see also Reid 2012.

[11] See Koyré 1957, 125–54; Gabbey 1990.

tions in the guise of Aristotelian science had been completely undermined by the discovery of the annual motion of the earth around the sun. The power which the Aristotelians granted to the celestial bodies must now be given to the immanent and omnipresent God (More 1681, 5–6). Such distaste for astrological thinking was shared by other Cambridge Platonists like Ralph Cudworth (1617–1688), who attacked not only the Stoic conception of astrological fate, but also such otherwise anti-astrological figures as Plotinus and Simplicius who thought the celestial bodies might still act as signs of future events (Cudworth 1678, 4–5).[12]

Butler started to answer some of these objections in the first book of his *Hagiastrologia*, before getting to More's chapters specifically in Book 2. When we look at all the wonder and variety in the generation of living things it goes without saying that they are the work of God, but the real question, Butler contended, is whether he operates with or without natural means. To hold the latter opinion is equivalent to saying that every act of generation is a new creation, which is absurd because there was only one Creation. Ever since then 'all things have come to pass by Nature, and therefore must there be some kind of Natural means for the production of all things', which 'means' must be a combination of earthly and celestial, seeing as the four elements alone cannot explain the wonder of natural operations (Butler 1680, 13). Butler described the now familiar metaphor of the astrological cosmos as a clock; wheels interlocking and circumscribing each other all the way up to the master-wheel or First Mover (Butler 1680, 16–17). As to whether or not there is a general Soul of the World mediating the action of the celestial bodies on the earth, Butler didn't seem to mind: *Utrum horum mavis accipe* ('Accept whichever you prefer') (Butler 1680, 14). He soon did pick a side, however, positing a universal vegetative soul in the heavens underpinning the celestial bodies with God's virtue. 'And thus immediately God ruleth in the Heavens', he concluded, 'and ruleth all the World mediately by the Heavens' (Butler 1680, 20).

So much for the basic metaphysical differences between the Cambridge Platonists and the astrologers. In the second chapter of his attack on astrology, Henry More took it upon himself to detail the main arguments used by the astrologers to justify their art, using as his primary sources David Origanus (1558–1628/29) and the Englishman Sir Christopher Heydon (1561–1623), in order then to refute them one by one. Firstly, More explained, the astrologers argue teleologically that all the stars and planets were made for more than just ornament. Secondly, they argue that the simple bodies of the elements cannot account for the perceived variety of natural phenomena. They also argue that if the sun is capable of such feats as spontaneously generating living creatures from dead matter unassisted, then surely the other celestial bodies have their own similarly powerful influences (More 1681, 39–41). In his defence, Butler declared himself quite satisfied with this chapter of More's. The would-be critic of astrology had done such a good job of explaining the arguments in favour of astrology that he had, so Butler thought, undermined the ambition of his work (Butler 1680, 41).

[12] On Cudworth and Stoicism, see Giglioni 2008; Sellars 2012.

It was in More's next chapter, where he took issue with each of these astrological pretences, that the cosmological and physical disagreements became apparent. He argued that we can safely assume that the fixed stars have no influence except for light and heat and not worry that they are thus useless:

> Because the later and wiser Philosophers have made them as so many Suns: which Hypothesis our Astrologers must confute before they can make good the force of their first Argument. And for the Planets, they have also suggested that they may have some such like use as our Earth has, i.e. to be the Mother of living Creatures (More 1681, 54).

In his crusade against astrological atheism, the assumption of ET life was one of More's strongest weapons. It answered the teleological challenge of the astrologers, retaining an office and influence for each body but localising and limiting them to heat, light, and the hosting of living creatures. We saw Thomas White take a similar approach (Sect. 5.3.2), and it was repeated after More by Cudworth, whose attack on astrology was met in the same work by praise of the recent improvements in astronomy and philosophy that have made every star a sun and supposed the existence of a plurality of habitable globes (Cudworth 1678, 882).

Butler's method of reply to this particular assault was to deny both the assumption and the consequence. The hypothesis that the stars are suns was 'a meer conceit', and if they were suns 'the more rather is to be expected from them', the sun being as powerful as it is (Butler 1680, 54–55). He was not convinced by More's vague allusions to higher beings making use of these suns, nor was he impressed by the possibility of living creatures on the other planets. He stuck fast to the belief that God made the planets in order to influence us here on earth, 'whatever *Chimera's* may dwell within them' (Butler 1680, 56). Henry More, of course, stuck to his own guns, and justifiably accused Butler of being ignorant of modern developments in astronomy, although his own talk of angels and ethereal genii composed of celestial matter inhabiting the vortices of the sun and fixed stars harked to the Cabbala rather than the telescope (More 1681, 76). As for the planets, if a secondary planet like the moon may be habitable, as the oceans and continents on its surface suggest, then there was even more reason to believe that the primary planets are all mothers of living creatures and so not made only for us. More concluded his rebuttal by declaring that the celestial bodies have 'manifold uses, partly in regard of others, and partly in regard of our selves, without Astrological Influence, and that therefore the force of this first general pretense of the Astrologians is defeated' (More 1681, 78).

The next proposition More took aim at was the claim that celestial influences are needed to explain the variety of terrestrial phenomena. He argued that even if you allow for some subtle differences between the material effects of the celestial bodies, these influences would still reach every point on the earth equally, and such material influences are not sufficient to explain terrestrial generation regardless. More instead credited his Stoic λόγος σπερματικός (spermatick principle)—identified with the Platonic world-soul—as being responsible for the moulding of matter 'into such shapes and virtues as its disposition makes toward' (More 1681, 55). Butler criticised what he saw as the arbitrary and opaque complexion of such a theory. Nature, he reminded us, does not act randomly but with order, and 'not according to any list or choice of her own', but by God's providential method (Butler 1680, 58).[13] She does not operate with only one instrument, but utilises the millions of tools available to her in the heavenly workshop. As to More's claim that the influence of each body reaches the entire surface of the earth equally, Butler pointed out More's own endorsement of the weapon salve, which, as we have seen, could act on a specific point from a great distance (Butler 1680, 59–60). On his part, More denied the veracity of this comparison. A logical and physical sympathy exists between a wound and the knife that made it. This sympathy cannot exist between the stars and embryonic matter, which is 'a pure Crystalline homogeneous liquor, as Dr. Harvey describes it, unvariegated of it self' (More 1681, 78–79).

More also took issue with the reasoning by which astrologers argue from the palpable influence of the moon to the supposed influences of her fellow planets. Butler, of course, defended this analogical reasoning in his vindication, without adding much substance (Butler 1680, 61). More's reply highlighted the division that the new astronomy had made between 'planet' and 'satellite', and so any analogy made from the moon could only apply to the moons around other primary planets, which, he accepted, should have similar effects on their planets to that of the earth's moon on it (More 1681, 83). Nor does the moon's effect on the tides justify a belief in occult influences:

> …the Laws of the *Aestus marinus* are executed sympathetically and synenergetically by the spirit of the World, and by the body of the Moon Mechanically as by his Instrument, and not by any strange Influence from her. And so the spirit of the World in Magnetical Phaenomenons acts Synenergetically and sympathetically from it self, but mechanically by those instruments of his operation, the Magnetick Particles which Cartesius calls the *particulae striatae* (More 1681, 84).

We saw hints of a similar theory in Gilbert, minus the particles, in which action at a distance is operated mechanically according to an essential sympathy underpinned by animate principles. Had More been so inclined, such a theory could easily be expanded to include planetary and stellar influences. Nevertheless, his loathing of astrological excess meant that he could not, while his long-standing partiality to pluralism meant he need not, do so. The stars supply heat and light to their own planets, and no significant amount thereof reaches the earth. The planets themselves are 'but heaps of dead matter much like that of the Earth' (More 1681, 60), and while Butler tried to flip this around by saying that such earthy planets would be more apt to influence our earthly human bodies (Butler 1680, 71), More would have none of it, allowing no action of the planets outside their own atmospheres (More 1681, 95–97).

This dispute between More and Butler is an example of how, by the latter stages of the seventeenth century, influence and inhabitation were increasingly in direct competition as teleological explanations in cosmology. Writing in 1680, Butler may seem to represent the 'old guard', but we should be mindful of letting our preoccupation with the anticipators of modernity blind us to the diversity of opinions at the

[13] In fact, More saw this spirit of nature as embodying the laws of nature, and so it did not act arbitrarily in the sense intended by Butler. See Henry 1990, 62–65; Harrison 2013, 139.

time. Pierre Borel's admission that 'we have much ado whom to believe' (Borel 1658, 6) well captures the state of seventeenth-century natural philosophy; pluralism still faced strong opposition on philosophical and theological grounds, and although astrology was being phased out of university curricula, it was still wedded to medicine and many other aspects of private and social life. More's reliance on a spirit of nature to explain everything from the generation of living creatures to the sympathetic foundations of magnetic and tidal influences shows the continuing difficulties that such phenomena presented to the new science. Newton's theory of gravity would take care of one problem, but generation and magnetism had to wait for the likes of Louis Pasteur (1822–1895) and Hans Christian Ørsted (1777–1851) to put them on a sound experimental footing.

The More/Butler argument was almost entirely at cross-purposes; a melange of insults and false-humility grounded upon analogies, suppositions and theological axioms. From a historical standpoint, it seems obvious to say that Henry More was the modern and Butler the traditionalist. We could even view the 'incommensurability' of their respective positions as evidence of a Kuhnian paradigm shift in progress. However, the only thing that really separated this dispute from similar ones conducted centuries before is that More was able to rebut the teleological argument for astrology with pluralism, whereas earlier critics had relied on timekeeping, navigation and the displayed glory of God as ways to rescue the utility of the celestial bodies without astrological influence. Butler's easy dismissal of More's pluralist theories as mere conceits shows how far telescopic observations and analogical thinking were from meeting the burden of proof. Astrology and pluralism were both conjectural theories with ancient pedigrees. As the revolution in astronomy approached fulfilment, it was only one of these theories—that concerning ET life—which was made commensurate with the increasingly orthodox world-system.

6.3 Pluralism Takes Hold

Neither Copernicanism, mechanism or pluralism were directly responsible for the marginalisation of astrology by the end of the seventeenth century. We have seen already the constructive role played by celestial influence in arguments for heliocentrism and ET life. On the other hand, the cases of White and More suggest that these could be seen as enabling factors in astrology's decline. For those who were opposed to judicial astrology for age-old reasons—that it had corrosive effects on politics, religion and the notion of free will—heliocentric cosmologies and mechanical philosophies could be interpreted so as to undermine the physical foundations of the art. Astrology, however, had strong teleological and providential functions. Some may have been happy to declare the unfathomability of God's purposes and leave it at that, but for others a plurality of inhabited worlds was an attractive alternative, and so a teleology of inhabitation could be asserted against one of influence.

An earlier example of this can be found in Claude Saumaise's (1588–1653) *De annis climactericis et antiqua astrologia diatribae* ('Learned Discussions on Climacteric Years and Ancient Astrology', 1648). In this voluminous attack on astrology and astrological practices, Saumaise takes the quite extreme approach of claiming that the assumption of celestial inhabitation precludes utterly any notions of celestial influence and its role in generation:

> Now, since the moon is a world and the other planets are worlds no less than the earth: as a world does not create a world, so none of the things that are generated in one world depend on the power of another world, by which they would be made better, larger, more animated or more lively. Everything obtains its qualities and the substantial form by which it exists from the natural character of its world where it is generated, not from another. … So, the moon or the other planets have no more regard for the origins and birth of the living beings of our globe, nor do they direct their acts, compose their fates or determine their ends, than this earth does with regard to the things that are born and live in the globe of the moon.[14]

Saumaise's denial of any mutual influence was unusual at the time, with most philosophers preferring some form of interconnection to a disparate collection of self-contained worlds, as we have seen. The endorsement of pluralism, however, would become more common in critiques of astrology later in the seventeenth century, just as attacks on astrology would become more common in discussions of the plurality of worlds.[15]

A few decades ago Michael Hunter brought to light an unpublished polemic against astrology written in 1673 by the future Astronomer Royal John Flamsteed (1646–1719).[16] Flamsteed had been an avid student of astrology in his younger days, but later turned against it, perhaps, Hunter suggests, due to his rejection or exclusion from the profitable almanac-publishing community (Hunter 1995, 254). It is important to note that again in this case, the motivations for an assault on astrology were not solely philosophical or empirical, and yet these were the tools which served for this purpose. The polemic tract, titled 'Hecker', was written as a preface to his ephemeris for the year 1674, and contains many of the usual ancient complaints against astrology as well as more recent ones concerning its incompatibility with a mechanical conception of nature. Inevitably, Flamsteed confronted the teleological refuge of the astrologer, who asks what use the stars will have if they are not

---

[14] Saumaise 1648, 775–76: 'Quandoquidem igitur Luna mundus est et caeteri planetae mundi haud minus quam haec terra, ut mundus mundum non creat, ita nec quidquam eorum quae in altero mundo gignuntur, ab alterius mundi potestate dependet quo meliora fiant, majora, vegetiora aut vivaciora. Ex sui mundi naturali proprietate unumquodque in eo gignentium accipit suas qualitates, et substantialem formam qua existit, non ab altero. … Non magis itaque Luna vel alii planetae nostri orbis animaliam originem ac genituram inspiciunt, actus dirigunt, fata componunt, finem determinant, quam haec terra eorum quae in orbe Lunae nascuntur et vivunt.' Translated in Vermij 2016, 309.

[15] Vermij argued that Saumaise's use of Copernicanism and pluralism against astrology is 'a very rare case, proposed at a time when the idea of celestial influence had already lost much of its credit' (Vermij 2016, 312), and thus the role of the new astronomy in the marginalisation of astrology should not be overstated (Hirai and Vermij 2017, 407). Hopefully this chapter will demonstrate that such arguments were not in fact so rare.

[16] References are thus to Hunter 1995.

allowed to act upon sublunary bodies. Flamsteed responded by alluding to the theory of 'Borellus' that the celestial bodies serve as habitats of other creatures.[17] He put the following challenge to the astrologer:

> Nay let him say, if the Caelestiall bodies are not far more excellently usefull, whilest they may, and probably doe receave and susteine creatures to give theire maker perpetual praises; then hee would willingly suppose them, whilest he argues that they operate, hee knows not how, upon our globe; and influence, not onely the great, but triviall affaires, thefts, rapines, & debaucheries of it (Hunter 1995, 281).

Flamsteed contrasted the probability, as he saw it, of ET life to the ignorant suppositions and trivialities of astrology. He tried to make clear that the pluralist route was a far more sensible and convincing answer to the teleological demands of astronomical cosmology. Flamsteed's attitude towards astrology was ambiguous, and the motivations behind this polemic were complex, but he was assured in that stance by the possibility of an alternative teleology of the heavens. His example is further evidence of the utility of pluralism as a tool to undermine the metaphysical and theological foundations of astrology.

Flamsteed's polemic was not published.[18] The reformed content of his ephemerides as well as the anti-astrological preface made it an ill fit for the marketplace of prediction-laden, astronomically-simple almanacs.[19] For the majority of the seventeenth century the average person in England engaged themselves with the heavens by consulting such almanacs, or by reading astrological political propaganda, or perhaps by consulting an astrologer or astrological physician. Talk of ET life, while gaining footholds through the works of Wilkins and Borel, as well as the lunar voyages of Godwin and Cyrano, was largely confined to the learned elite in the coffee shops, or to vague and enigmatic mentions in dense cosmological publications such as we have dealt with in this discussion. This state of affairs was altered somewhat by the publication of two works: *Entretiens sur la pluralité des mondes* ('Conversations on the Plurality of Worlds', 1686) by Bernard de Fontenelle (1657–1757) and *Κοσμοθεωρος, sive de terris coelestibus, earumque ornatu, conjecturae* ('Cosmotheoros, or Conjectures on the Celestial Earths and their Furniture', 1698) by Christiaan Huygens (1629–1695).[20] These works—the first an imagined conversation between the author and a marchioness over the course of several evenings, the second a series of theologically guided conjectures about life on the other planets—presented the plurality of worlds philosophy to a wider audience, and both were translated into English within a year.[21] Fontenelle's work was translated into

[17] Hunter identifies Borellus as Giovanni Alfonso Borelli (1608–1679), but considering the subject matter and the fact that the following sentence refers to the Cartesian philosophy, it is more likely Pierre Borel (1620–1671).

[18] For suggestions of reasons why not, see Hunter 1995, 261–66.

[19] See Capp 1979.

[20] On Fontenelle, see Lovejoy 1948, 130–33; Dick 1982, 123–40; Crowe 1986, 18–21. On Huygens, see Dick 1982, 127–35; Crowe 1986, 120–22.

[21] References given here are to later editions: Huygens 1722; Fontenelle 1809.

English five separate times, and went through 33 French editions (Dick 1982, 136–38).

The contents of these works have been summarised and analysed extensively elsewhere, and so only a few comments will be made here. They both describe a universe made up of a network of sun-centred Cartesian vortices filled with habitable planets, although there are important and telling differences between their conjectures about the planetary inhabitants. Their popularity marked a new beginning for pluralism as a mainstream subject, and makes compelling the argument that Cartesian rather than Brunian philosophy had the largest impact on its growing acceptance (Lovejoy 1948, 124–25; Dick 1982, 126). Huygens' strong astronomical pedigree granted a level of scientific respectability, while the simplified theoretical content and the romantic, conversational style of Fontenelle allowed readers to bypass the technicalities and 'jump directly to Enlightenment', as Westman puts it (Westman 2011, 502). This is, of course, a loaded statement, but it is indicative of how we still today perceive the shift from an enclosed, anthropocentric cosmos to the unbounded and inhabited universe of the Enlightenment as an intellectual and civilisational forward leap.

The particularly unscientific nature of the argumentation in these two works has been commented on many times.[22] The extent of observational and physical knowledge of the time was not sufficient to reach conclusions by an appropriate scientific method. Indeed, Fontenelle himself seemed to acknowledge as much. When, in one of their night time conversations, the author declares to his companion that the moon may be furnished with everything it needs to sustain life, the Marchioness replies: 'That is to say… you know all is very well, without knowing how it is so; which is a great deal of ignorance, founded upon a very little knowledge' (Fontenelle 1809, 59). The author is naturally unperturbed, having full faith in the legitimacy of his analogical reasoning. Astrology is not discussed at all in the week's conversations, perhaps purposefully. In his summary of reasons to believe that the planets are inhabited, Fontenelle didn't even consider astrology worth a mention:

> …you have all the proofs you could desire in our world. The entire resemblance of the planets with the earth, which is inhabited, *the impossibility of conceiving any other use for which they were created*, the fecundity and magnificence of nature, the certain regards she seems to have had to the necessities of their inhabitants, as in giving moons to those planets remote from the sun, and more moons still to those yet remote; and what is still very material, there are all things to be said on one side, and nothing on the other; and you cannot comprehend the least subject for a doubt, unless you will take the eyes and understanding of the vulgar (Fontenelle 1809, 129–30. Emphasis added).

That astrology formed part of this 'understanding of the vulgar' was hinted at in an earlier work of Fontenelle's, his *Nouveaux dialogues des morts* ('New Dialogues of the Dead', 1683). In an imagined dialogue between the deceased Joanna I of Naples and her astrologer Anselm, the queen asks that he provide her a prediction about her future.[23] He claims that he cannot, and says further that such anxiety about

[22] See, for example, Jaki 1978, 55. It is also one of the main arguments of Crowe 1986.

[23] This is Dialogue 6 in the third section containing dialogues of the moderns: Fontenelle 1683, I, 247–63.

the future is a feature of the living, not the dead. 'The future is the great lure of mankind', he says, 'and we astrologers know this better than anyone.'[24] Astrologers play on this desire for future knowledge and tell stories of zodiacal signs and planets that are hot or cold, male or female, good or bad. The queen points out the irony that the man who was once her astrologer is now speaking ill of the art. 'Listen', replies Anselm, 'for a dead man does not like to lie. In truth, I deceived you with this astrology that you esteem so highly.'[25]

Huygens was more direct than Fontenelle in his juxtaposition of ET life and astrology. At the start of Book 2 of the *Cosmotheoros*, he described reading Kircher's *Itinerarium exstaticum* and being amazed by the 'heap of idle unreasonable stuff' he found written therein (Huygens 1722, 101). He didn't quite understand how someone could describe a universe made up of multiple solar systems with earth-like planets and still maintain that the earth is the only abode of life:

> Since most of these Worlds are out of the Reach of any Man's sight, as he owns they are, I cannot think for what purpose he makes so many Suns to shine upon desolate Lands (like our Earth in everything, he says, only that they have neither Plants nor Animals) where there's no one to whom they should give light. And from hence he still falls into more and more Absurdities. And because he could find no other use of the Planets, even in our System, he is forced to beg help of the astrologers; and would have all those vast bodies made upon no other account than that the whole universe might be preserved and continue secure by their means, and that they might govern the mind of man by their various and regular influences (Huygens 1722, 103).

Huygens may have found Kircher's reasoning absurd, but his own methodology was hardly more 'scientific'. He criticised the way that Kircher tried to gratify astrology by painting the physical conditions of each planet in line with their astrological quality. Venus is a pleasant place, Mercury airy and brisk, Saturn is melancholic and dark. 'All this and such like Stuff his Genius teaches him', remarked Huygens wryly (Huygens 1722, 104), although the same scepticism was not apparent in his own conclusion that each planet has 'Fields warm'd by the kindly Heat of the Sun, and water'd with fruitful Dews and Showers' (Huygens 1722, 28–29). Huygens' universe has every planet filled with plants, animals and rational creatures, all functioning pretty much like they do on earth. His vision was quite drastic in its lack of anthropocentrism, based on the premise that the earth should have no advantage at all over any of the other planets, but it was also anthropo-normative, assuming that everything on earth, from the growth of plants in the soil to the practice of astronomy by upright-standing rational animals, is repeated throughout the solar system (Huygens 1722, 38, 61–63).

Fontenelle and Huygens also show us how a belief in ET life began to take over some of the functions of astrology beyond just that of teleology. Both works promoted pluralism as a way of encouraging a cosmic perspective. Advocates for astrobiology outreach do much the same today, and the form of the argument hasn't

[24] Fontenelle 1683, I, 252–53: 'le grand leurre des Hommes, c'est toûjours l'avenir, et nous autres Astrologues nous le sçavons mieux que personne'.

[25] Fontenelle 1683, I, 254: 'Ecoûtez; un Mort ne voudroit pas mentir. Franchement, je vous trompois avec cette Astrologie que vous estimez tant.'

changed much since Scipio had his dream in the first century BCE.[26] Just as visions of flights through the celestial spheres once did, a reasoned awareness of the multiplicity of inhabited worlds simultaneously makes a person disdainful of worldly success and grateful for the amenities that God has provided for his creatures (Huygens 1722, 10–11). These works on pluralism also demonstrate how the belief provided an incentive to further develop the astronomical sciences. Instead of improving theorics and planetary tables in order to make more accurate predictions, Fontenelle and Huygens looked forward to improved telescopes and perhaps even space-travel in order to find out more about our celestial neighbours.

The supposition of life on the other planets also began to give a new shape, order and logic to the solar system in a way which had been missing ever since the sun took the earth's place in the centre. Both Fontenelle and Huygens postulated differences between the denizens of each planet based on the distance of the planet from the sun. For Huygens, this difference was 'not so much in their Form and Shape, as in their Matter and Contexture' (Huygens 1722, 24). According to Fontenelle, the inhabitants of Mercury are vivid and nimble, while those of Saturn are slow and deliberate. We here on earth, being situated in the middle between two extremes, are 'a mixture of the several kinds which are found in the rest of the planets' (Fontenelle 1809, 98–99). Humanity, rather than being the centre of the universe, became its measure—for Huygens we are the median; for Fontenelle the mean. Astrology, which had once reigned as the predominant cosmological paradigm and permeated all society, was by the end of the seventeenth century a minority interest, at least among the intellectual elite (Wright 1975, 404–5). Pluralism, while in no way solely responsible, at least played a contributory and enabling role in this marginalisation. In the final section, we will look at Newtonianism and Natural Theology, exploring how astrology was side-lined within a metaphysics that stressed God's immediate action in the world, while the debate over ET life was framed in increasingly religious, non-naturalistic terms.

6.4 Natural Theology and the De-astrologisation of Generation and Providence

'In eighteenth-century Newtonian natural philosophy, the plurality of worlds tradition found its culmination and its triumph' (Dick 1982, 175). This was the assessment of Dick, and while its veracity depends on the weight you give to high-end natural philosophy in comparison to other cultural forms, it seems fair to say, with Crowe, that the combination of Newtonianism, Christianity and pluralism 'would within a few decades become commonplace' (Crowe 1986, 24). The vehicle for this combination was natural theology (often called physico-theology), a project which used knowledge of the natural world to construct a design and providential

[26] As just one example, see Crawford 2018.

argument for God. In this regard, it was not a new enterprise, and instructive comparisons can be drawn to the similar endeavours of Thomas Aquinas or Philip Melanchthon discussed in Chap. 2. The key difference with respect to our discussion is that the natural theologians around the turn of the eighteenth century de-emphasised the role of celestial influence in the process of generation and in God's providential governance of the universe. The creation of living creatures in the first instance, as well as the continuity of species, was granted more directly to God's design, rather than to any secondary instruments or mechanical operation of matter, and the possibility of any consequent spontaneous generation of life was denied. The function of celestial influence as an explanation of 'what the celestial bodies are for' was increasingly replaced by pluralism—God's providence was demonstrated by the stable maintenance of a plurality of inhabited earth-like bodies, rather than by the maintenance of this earth via the combined operations of the celestial realm.

This reversed the dynamic between astrology and pluralism that had largely held from late antiquity to the Renaissance. We mentioned in Chap. 2 how Manilius' astrological poem had opposed an ordered, unified and providential cosmos to the chaotic, mechanistic pluralism of Lucretius' atomist verse (see Sect. 2.1). Throughout the Middle Ages and into the sixteenth century, celestial influence played a key role in support of a designed universe and in arguments against the perceived evils of Epicureanism. In the seventeenth century, astrology was seen by many as naturalistic, deistic and deterministic; attributes that associated it with the atheistic mechanism and materialism of Hobbes and Baruch Spinoza (1632–1677).[27] Indeed, Hobbes' physical defence of celestial influence in his critique of White's *De mundo* illustrates this connection. One solution to this was to take the atoms and multiple worlds of Epicurus and substitute divine governance for chance.[28] In this context, the assertion that the celestial bodies were created to provide habitation for other creatures obviated the need to grant them power over the sublunary realm, and undermined the teleological foundation of 'astrological atheism'.

Peter Harrison's suggestion, as mentioned in the introduction (Sect. 1.2), was that the decline of astrological prediction and the associated belief in celestial influences made the problem of explaining the utility of the celestial bodies more acute. The following discussion will suggest a revision to that assessment. Many natural theologians were explicitly critical of an astrological tradition that was still prevalent in their society, and so they, like White, More, Flamsteed and Huygens, deliberately championed a principle of inhabitation in opposition to one of influence. The natural philosopher Thomas Burnet (c. 1635–1715), for example, in his *Telluris theoria sacra* ('Sacred Theory of the Earth', 1681), expanded his naturalistic interpretation of the Biblical Flood and the future Conflagration to apply likewise to the other inhabited planets (Burnet 1697, I, 113–14). The second part of that work,

[27] There is a large body of literature on Spinoza, but one more relevant article in this context is Simonutti 2001.

[28] On the appropriation of Epicurus by seventeenth-century English anti-atheists, see Sheppard 2015, 102–25. Sheppard's otherwise thorough analysis of anti-atheism in this period does not discuss pluralism or astrology. See also Clucas 1991.

published in a new edition of 1689, criticised the opinion of some 'astronomers' that the end of the world could be brought about by a particular configuration of the heavens. 'Pray what reason can you give', he asked, 'why the Planets, when they meet, should plot together, to set on Fire their Fellow-Planet, the Earth, who never did them any harm?' (Burnet 1697, III, 19). The ensuing censure of astrology claimed that the idolatrous origin of astrological beliefs had been undermined by the realisation that they are opaque terraqueous globes like the earth, and Burnet declared that it was high time 'to sweep away these cobwebs of superstition, these reliques of Paganism' (Burnet 1697, III, 19).

One of the most influential works for the Newtonian iteration of natural theology was *The Wisdom of God Manifested in the Works of the Creation* (1691) by the theologian and naturalist John Ray (1627–1705). His separation of the design argument into two sections—one arguing from the structure of the heavens, the other from the structure of living organisms—would be copied by many later authors. In his discussion of the heavens, Ray argued that the planets are most likely furnished with as great a variety of corporeal creatures as the earth (Ray 1714, 18–19). The moon as well, while exerting useful influences upon the earth and supposedly affecting the growth of plants and animals, has a further use in maintaining the creatures 'which in all likelyhood breed and inhabit there', on which subject he referred the reader to the works of Wilkins and Fontenelle (Ray 1714, 66–67). Concluding his discussion of the utility of the celestial bodies, he mentioned that even eclipses of the sun and moon can be of great use to 'knowing Men', even though 'they be frightful Things to the superstitious Vulgar, and of ill Influence on Mankind, if we may believe the no less superstitious astrologers' (Ray 1714, 68).

This distaste for astrological thinking is evident throughout Ray's botanical work as well.[29] In *The Wisdom of God*, he attempted to demolish a doctrine that historically had been central to the relationship between biology and celestial influence: equivocal or spontaneous generation (Ray 1714, 300–326).[30] His metaphysical or theological argument against this was that to bring forth a living creature out of inert matter requires an omnipotent agent, and so spontaneous generation is akin to Creation. For observational and empirical evidence to support this, he was able to cite the experiments of Marcello Malpighi (1628–1694), Francesco Redi (1626–1697) and Antonie van Leeuwenhoek (1632–1723), all of which seemed to suggest that instances of supposedly spontaneous generation were actually the result of invisible or unnoticed eggs or seeds.[31] If equivocal generation could be disproved, Ray thought, then it would take away one of the main supports of atheism. 'They cannot then exemplify their foolish Hypothesis of the Generation of Man and other

[29] See Raven 1986, 98.

[30] Ray had long been sceptical about spontaneous generation, and his strong assertion in the first edition of *The Wisdom of God* that there was no such thing drew an attack by an anonymous correspondent. This led to Ray expanding his refutation in the second and subsequent editions. See Raven 1986, 375.

[31] See Farley 1977, 14–18.

Animals at first,' he argued, 'by the Like of Frogs and Insects at this present Day' (Ray 1714, 322).

It was exactly this objective—the combatting of atheism—that motivated Robert Boyle (1627–1691) to bequeath a certain sum of money for an annual series of public lectures. Boyle was quite enthusiastic about the benefits of a teleological consideration of nature for encouraging a belief in God, but he was unsure whether astronomy was the most appropriate field of study in this regard. In his *A Disquisition about the Final Causes of Natural Things* (1688), he suggested that the benefits which the celestial bodies bestowed on the earth did not rule out the possibility of other ends, but he urged discretion in light of our lack of exact knowledge about the system of the heavens (Boyle 1688, 235–36). The bodies of animals and plants, he argued, could provide more 'clear and cogent arguments' for the wisdom and design of God (Boyle 1688, 43). In spite of this, the Boyle Lectures would become instrumental in the replacement of a celestial teleology of influence with one of inhabitation, thanks in part to the relationship between the first Boyle Lecturer, Richard Bentley (1662–1742), and Isaac Newton.

Newton himself said little about either ET life or astrology. There is a crossed-out passage in his manuscripts which asks why all the immense spaces of the heavens should be 'incapable of inhabitants'. This remark was published, without reference to its deletion, by David Brewster (1781–1868), who invoked the support of Newton for his own pluralist theories (Brewster 1855, II, 354). This possible private or tacit acceptance of pluralism was accompanied by a probable rejection of astrology, although this too was not discussed openly or at length.[32] Bentley, however, engaged directly and publicly with both these subjects. A classicist by training, it has been suggested that Bentley first contacted Newton because he was working on an edition of Manilius' *Astronomica* (Guerlac and Jacob 1969, 314–15).[33] He was assisted in this endeavour by Edward Sherburne, whose English translation, with its abundant notes and long scientific appendices, was undoubtedly a major inspiration for Bentley. Newton was included as a worthy contemporary in the catalogue of astronomers that Sherburne appended to his translation, and thus it may have been that Bentley approached Newton with a view to similarly complement his Manilius with a modern astronomical perspective.

It is also possible that Newton played a part in selecting Bentley to deliver the inaugural Boyle Lectures, although other trustees were familiar with the young scholar.[34] The eight lectures, or sermons, that he gave in 1692 were printed in that and the following year as *A Confutation of Atheism* in three parts. The first part was (meta)physical, arguing, *à la* Digby and More, that the limitations of matter necessitated the existence of the soul. The second and third parts mirrored Ray's biology/astronomy divide, arguing against atheism from the origin and structure firstly of the human body and then of the larger universe. When it came to opinions about the

[32] On Newton and astrology, see Schaffer 1987; Rutkin 2006.

[33] It is tempting to see this as evidence for an early interest in astrology, but Bentley's motivations for the edition were probably more philological than scientific. See Haugen 2011, 212–13.

[34] On Bentley's relationship with Newton in regard to pluralism, see Dick 1982, 144–49.

origin of mankind, Bentley divided the atheists into four rough categories: Aristotelian, Epicurean, astrological and mechanical. The first denies that the human race had a beginning, arguing that there has been an eternal succession of generations. The second group invokes a chance concourse of atoms. It is the latter two brands of atheism which concern us here.

The 'modern astrological atheist', as Bentley described him, credits a particular extraordinary conjunction or aspect for the initial generation of humanity (Bentley 1836–1838, III, 62, 71). To expurgate this notion, Bentley took on the entire astrological tradition, arguing that its main theoretical pillars—geocentrism and ensouled planets—had been obliterated (Bentley 1836–1838, III, 67–72). He could forgive 'some great men of the last age' for being addicted to astrology (and here we might think of Manilius), but maintained that 'at this time of day, when all the general powers and capacities of matter are so clearly understood, he must be very ridiculous himself that doth not deride and explode the antiquated folly' (Bentley 1836–1838, III, 71). It was an unwelcome irony, he thought, that an art which was historically reliant upon idolatry had become 'the tottering sanctuary of Atheism' (Bentley 1836–1838, III, 71).

Satisfied with his confutation of an astrological genesis of humanity, he then moved on to the mechanical theory. This also relies on celestial influence, albeit of a more general kind, ascribing the first generation of man to the action of the sun upon 'duly prepared matter' (Bentley 1836–1838, III, 62). His approach to this problem involved a refutation of exactly that kind of generation that we saw described by Digby (Bentley 1836–1838, III, 80–84). Even theories like that of the Cambridge Platonists involving the action of a plastic spirit of nature were unacceptable. The only invocation of nature by Bentley was in regard to the 'settled method' of the propagation of mankind—and, by extension, other creatures—following an initial direct act of creation. He thus argued similarly to Ray that nothing had given more support to atheism than 'this unfortunate mistake about the equivocal generation of insects' (Bentley 1836–1838, III, 85).[35]

There is an interesting letter from Bentley to Edward Bernard (1638–1697) which reveals some of the thinking behind the tactics of his refutation.[36] Bernard, it seems, had suggested that a better target for Bentley's sermons might have been deists or Jews, rather than atheists, who had few books published. Bentley denied that they have no written books, arguing that 'not one English Infidel in a hundred is any other than a Hobbist' (Wordsworth 1842, 39). Bernard also suggested that the idea of an astrological origin of mankind was Bentley's own invention, and that no one ever believed it. Bentley's reply is particularly germane:

> But 'tis your happiness, that you have not known by conversation what Monsters of Men have been of late days. You know the grounds of the old ones, that derived us out of the soil from mechanism or Chance, was, that equivocal generation of frogs and insects, and plants *sine semine*. So that they said, when the earth was fresh and vigorous the more perfect animals were produced out of her. Now, therefore, because the generations of Plants and

[35] See also Goodrum 2002.

[36] Richard Bentley to Edward Bernard (May 28th, 1692), no. 19 in Wordsworth 1842, 38–41.

> insects are reduced to the starry influences, they carry in consequence the production of ourselves to the same Cause. Besides Cardan, Caesalpinus, and Berigardus, etc. do in express words ascribe it to planetary influences: and 'tis now the reigning opinion of the most learned living Atheists among us; and therefore ought not to be past by (Wordsworth 1842, 40).[37]

Bentley was of course right in saying that there were astrological theories about the origin of mankind, as indeed there were about the origin or creation of the world itself.[38] We need just look to the sources that Bentley himself singled out to discover what troubled him.

The sixteenth-century Pisan philosopher and botanist Andreas Cesalpino (1519–1603) discussed this subject in his *Quaestionum peripateticarum libri V* ('Five Books of Peripatetical Questions', 1571), under the purview of the following question: 'Whatever is produced from a seed can likewise be produced without a seed'.[39] This chapter contains an example of exactly that Averroist, astrologised Aristotelianism discussed in Chap. 2. Cesalpino referenced Aristotle for the theory that humans and quadrupeds might originally have been born from the earth (*GA*, III.11, 762b28–30) and also that in nature, as in art, things that are produced as the result of some faculty can also occur by accident (*Metaphysics*, VII.7). Combining these with Aristotle's theory that terrestrial change proceeds from the heavens as an efficient cause (*GA*, II.2–3; *Meteorology*, I.3), Cesalpino arrived at the following conclusion:

> We therefore conclude from these principles that all things were indeed produced in the beginning by intelligence as the first mover and by the heavens as an instrument, after which, those things that had obtained a more perfect nature were produced by turns from each other.[40]

This is the method of reasoning that Bentley criticised, arguing from the role of the heavens in ongoing spontaneous generation to suggest an original equivocal generation of more perfect animals, including humans, brought about by the celestial bodies.[41]

Cesalpino—along with Berigardus, Cardano, Vanini and the mechanical philosophers—had been the subject of an earlier assault against atheism by Samuel Parker (1640–1688), which may have been Bentley's source (Parker 1678). Whether this

[37] The figures referred to are Girolamo Cardano (1501–1576), Andrea Cesalpino (1519–1603) and Claudius Berigardus (1578–1663).

[38] The idea of a *thema mundi*, a horoscope showing the positions of the planets at the beginning of the world, has a long tradition. The idea of a particular astrological configuration being responsible for the first creation of man has a somewhat different lineage, seemingly derived from Zoroastrianism, but being transmitted via the Pseudo-Aristotelian Hermetica. See Burnett 1997, 41–42; Raffaelli 2001; De Callatay and Saif 2017.

[39] 'Quaecunque ex semine fiunt, eadem fieri posse sine semine', in Cesalpino 1593, fols 104ᵛ–109ᵛ.

[40] Cesalpino 1593, fol. 109ᵛ: 'Concludimus igitur ab his principiis, ab intelligentia quidem tanquam primo movente, a coelo autem tanquam instrumento: omnia quae hic sunt oriri primo, secundario autem a se invicem in iis quae perfectiorem naturam adepta sunt.'

[41] It should be mentioned that Cesalpino excluded the production of souls, human or otherwise, from this process. See Cesalpino 1593, fol. 104ᵛ.

was the 'reigning opinion' among atheists of Bentley's time is another matter, but there were similar iterations of this theory close to home. Burnet, for example, had suggested that the ether or the heavens could have supplied the 'male' influence in a spontaneous generation of plants and animals (not men) at the beginning of the world, citing Aristotle's comparison of semen and *pneuma* in support (Burnet 1697, II, 135–36). Here we see the conflation, which Bentley recognised, of astrological and mechanical or Lucretian theories of the origins of life.[42] The combined threat of these atheistic notions led Bentley to refute not only an astrological origin of man but also to deny the reality of spontaneous generation, which seemed to support the idea that 'starry influences' have the ability to produce life out of non-living matter—in short, it led him to de-astrologise generation.

In turning to the larger system of the universe, and specifically its origin, Bentley reassured the reader that the astrological atheists would not pose a problem, because the planets and zodiac signs could not logically be the cause of their own existence (Bentley 1836–1838, III, 131). The main targets, therefore, were the eternal world of the Aristotelians, the infinite chaos of the Epicureans, and the naturalism or deism of the mechanical philosophers. It is the Newtonian theory of gravity that takes centre stage in these discussions, not only in its power but in its limits. Gravity alone, Bentley argued, could not have produced the current system of stars and planets out of an initial chaos of matter. Moreover, the continued operation of gravity relies immediately on God himself. As a force applicable to celestial motion, Newton's theory of gravity was a development of Kepler's celestial physics and Gilbert's magnetic theory. Gilbert described magnetism as the earth's astral virtue and believed it to be inherent to the earth's substance, and Bentley's description of Newtonian gravitation, while comparable, contains an important difference. He described gravity as 'an operation, or virtue, or influence of distant bodies upon each other through an empty interval, without any effluvia, or exhalations, or other corporeal medium to convey and transmit it'. So far this looks similar to Gilbert's effused forms, but Bentley concluded from these properties that this power 'cannot be innate and essential to matter', but is directly impressed and maintained by the divine power (Bentley 1836–1838, III, 163).

There is no time nor need here to delve into the intricacies of the debate about the nature and operation of gravity, but it is worth thinking about how this theory, much more so than the otherwise similar ideas of Gilbert or Otto von Guericke, de-astrologises the interaction between the celestial bodies. The order and motion of the heavens is not maintained mechanically or by any other variety of secondary natural cause, but rather by the immediate power of God. It is in light of these de-astrologising tendencies in Bentley's sermons that we should view his enthusiasm for ET life. Discussing the ends and final causes of these higher, inanimate parts of the world, he pointed out the usefulness of some for humanity, but quickly concluded that most could not have been created for our sakes. His approach to pluralism and the nature of ET life is actually strikingly similar to that of Thomas White (see Sect. 5.3.2). Bentley stated that 'all bodies were formed for the sake of intelli-

[42] See Poole 2015.

gent minds', and so the other planets serve as habitats for other rational creatures as the earth does for men. Theological concerns about the Fall or the Incarnation are unfounded, because the fact that these planetary inhabitants may be rational does not mean they are 'men'. As White had done, Bentley defined humanity not merely as the rational animal, but as a species within that genus, differentiated by its unique sub-class of soul and the particular structure of its organic body (Bentley 1836–1838, III, 175–76). After Bentley, this would become a common way of approaching the question of ET life.[43]

Bentley didn't give any specific details about these other potential species, but he did feel the need to address the possible objection that the inner- and outermost planets of our system would be too hot and cold respectively to support life. Part of his response is quite revealing: 'the laws of vegetation, and life, and sustenance, and propagation, are the arbitrary pleasure of God, and may vary in all planets according to the divine appointment' (Bentley 1836–1838, III, 182). This conviction, which separates the question of ET life from a belief in the uniformity of nature, is linked to the issue of voluntarist versus intellectualist theology. Newtonian natural philosophy was characterised by a theological approach which prioritised God's will, as opposed to the approach which prioritised his intellect, usually associated with the continental mechanical philosophies.[44] These theological differences were central to the issue of natural laws.[45] An intellectualist might have felt justified in approaching this subject *a priori*, on the basis that God is restricted by certain logical and natural necessities in the creation and continued maintenance of the world. A voluntarist, on the other hand, considered God to be completely free in all his actions, and so the only way of discovering the method and laws that he did and *does* use was to do it *a posteriori*, through the accumulation of empirical and observational data.

Bentley's approach to pluralism was not intellectualist, but neither was it empirical. By invoking the arbitrary pleasure of God in this instance, he was effectively saying that life on earth provides no basis for extrapolating more general biological laws of nature. His belief in ET life had nothing to do with the universality of matter and the forces inherent to it or acting upon it; it was almost purely teleological. Herein lies the difficulty. Bentley's voluntarist theology should, in theory, have prevented him from making conclusions about God's actions in areas beyond the limits of empirical observation. In the case of ET life, it blatantly did not.[46] His fear of astrology as a threat to religious belief may provide a clue in this regard. Astrological and mechanical explanations, in both celestial and terrestrial matters, granted too much power to nature at the expense of God, which Bentley saw as effectively athe-

[43] See Jenkin 1700, II, 223; Sturmy 1711, 25. The theological issues that arise from the question of ET life are dealt with extensively in Dick 1982. See also Almond 2006.

[44] There is a scholarly disagreement about how helpful this dichotomy is in understanding early modern natural philosophy. See Harrison 2002; Harrison 2009; Henry 2009.

[45] For the above authors' views on this derived issue, see Henry 2004; Harrison 2013.

[46] This might support Henry's claim that early modern natural philosophers didn't commit themselves exclusively to either voluntarism or intellectualism, but were more pragmatic, 'cutting their cloth to suit the prevailing conditions'. See Henry 2009, 82.

istic. The replacement of celestial influence with celestial inhabitation and the stated dependence of life upon the arbitrary agency of God allowed Bentley to present a vision of the cosmos that kept both opponents at bay.

In Bentley's published sermons, Boyle's warning about the teleological limits of astronomical investigations was ignored, and the young Newtonian filled the heavenly bodies with innumerable instances of ET life. Along with Fontenelle and Huygens, Bentley is an indication not only of how inhabitation was replacing influence as the answer to the question 'what are the celestial bodies for', but also of how pluralism was replacing astrology as the speculative or conjectural side of astronomy. It is important to note, however, that natural theology was not a homogeneous or static discipline. Certain writers were less favourable than Bentley towards pluralism, while others were not as reticent about celestial influence. There are many varied and interesting theories on display in the works of natural theology in this period, but as this book reaches its conclusion there is only space to mention a few.

The physician George Cheyne (1671/2–1743) endorsed pluralism based on the analogy between the earth and the planets and satellites, and, like Huygens, belittled an astrological interpretation of the utility of those bodies. He found it impossible to conceive that they were made 'for no other use but to twinkle to us in Winter Evenings, and by their Aspects to forebode what little Changes of Weather, or other pitiful Accidents were to be expected below' (Cheyne 1705, 110). The botanist and physician Nehemiah Grew (bap. 1641–d. 1712) also thought there was good reason to believe that the moon and the planets were inhabited, and that the fixed stars were other suns (Grew 1701, 10). He allowed more room than Bentley, however, for the role of secondary instruments in the administration of nature. One dimension of secondary causation was provided by the dominion of the celestial bodies, 'both the Planets, and fixed Stars; but chiefly, the Sun and Moon', over the earth. The subtlety of the intervening ether allowed for the transfer of light, heat and 'other qualities' (Grew 1701, 87–88).

Robert Jenkin (bap. 1656–d. 1727) demonstrated less enthusiasm about inhabited planets in his work entitled *The Reasonableness and Certainty of the Christian Religion* (1696). 'The Stars may be of great Benefit and Usefulness in the World,' he suggested, 'tho' they neither have that Influence which Astrologers vainly suppose, nor are as Suns to other Earths' (Jenkin 1700, II, 219). His attempts to demonstrate this usefulness highlight some of the remaining possibilities for a teleology of influence within a Newtonian system. The first use that the stars have, he argued, is to 'keep the circumjacent Air or Aether in Motion', preventing it from stagnating, and maintaining 'that perpetual Circulation of Fluid Matter, which passes from Orb to Orb, through the Universe, and gives Life to all Things' (Jenkin 1700, II, 219–20). So the motion of the celestial bodies propels the circulation of a Stoic *pneuma* that has vital properties. At the same time, their gravitational effects keep the different parts of the universe in equilibrium, 'by Mr. Newton's principles' (Jenkin 1700, II, 222). A passage added to later editions of the work expanded on more possibilities. According to these same principles of gravitation, the conjunction and opposition of planets should have some effect on the earth. Comets, meanwhile, cannot be inhabited because they approach too close to the sun, and so their use must in some way involve influence (Jenkin 1721, II, 229).

Cometography was, in general, the main area where astrological influence persisted within the Newtonian system. Comets were naturalised by the demonstration that they are orbital bodies whose return can be predicted. Their highly eccentric orbits, however, and their uncertain material constitution, made it difficult to say exactly how they fit, teleologically speaking, into the celestial schematic, and so discussions about their role continued to include theories of influence. Newton's own tentative theory was that comets restocked planets with fluid matter that they lost over time (Kubrin 1967, 336; Schaffer 1987). Jenkin claimed that the philosophers of his time agreed 'that the vast Quantity of Moisture in the Atmosphere of Comets, must cause Changes in our Air' (Jenkin 1721, II, 229). They may be a natural means of executing God's providence, in the form of either blessings or punishments, and therefore planets may well serve the same purpose, obviating the need for them to be inhabited.[47]

Daniel Sturmy (d. 1722) took up many of Jenkin's objections in his own work on pluralism. He conceded that other orbs may influence our own in unknown ways 'by Pipes and Channels, of very subtile Matter', but rejected the implication that we should therefore limit them to human uses (Sturmy 1711, 17–18). He copied Jenkin word-for-word in places, such as in the case of the stars serving to keep the ether in motion, or of planetary satellites providing a possible method for determining longitude. These 'wise ends', however, hold true whether the planets are inhabited or not, and so rather than contracting our contemplation of providence, we should favour the option which enlarges our admiration of God (Sturmy 1711, 26–27). Crowe has argued that isolated voices of opposition, like that of Jenkin, 'had little effect in stemming the ever increasing enthusiasm for the pluralist message' (Crowe 1986, 36). This message could still accommodate ideas of celestial influence, it seems, but an astronomical teleology built solely on influence was increasingly untenable.

One of the more important natural theologians of the early-eighteenth century was William Derham (1657–1735), who, like Bentley, was a Boyle Lecturer. His *Physico-theology* (1713), arising out of the lectures he delivered at St Mary-le-Bow, London, throughout 1711 and 1712, contained a declaration that spontaneous generation was 'a Doctrine so generally exploded', that there was no longer a need to disprove it (Derham 1720, 244). This work on terrestrial natural theology was followed up by *Astro-theology* (1715), dealing, as one might suspect, with celestial bodies and phenomena. Both of these works went through many editions in the eighteenth century, and were translated into several different languages. Just as the question of equivocal generation was considered settled, so too, to a certain extent, was the question of pluralism. *Astro-theology* begins with a preliminary discourse on the world-system and the habitability of the planets intended as an aid to the

[47] Later philosophers would propose that comets were also inhabited, most notably Johann Heinrich Lambert (1728–1777) in his *Cosmologische Briefe über die Einrichtung des Weltbaues*. Lambert 1761, 31–58, translated in Lambert 1976, 68–80.

reading of the main work. The upholders of the new system of the world, Derham maintained, conclude that 'all the Planets of the Sun and of the Fixt Stars also' are habitable worlds, 'places, as accommodated for Habitation, so stocked with proper Inhabitants' (Derham 1731, liii–liv).

Derham also dealt with the question of celestial influences. What exactly these influences were, he thought, was difficult to determine, 'although vainly pretended unto by the judicial Astrologers' (Derham 1731, 184). His concern with astrological atheism, however, doesn't seem as great as Bentley's. In Derham's *Astro-theology* we see semblances of the symbiosis of influence and inhabitation characteristic of the previous century. He described 'an admirable Oeconomy observable throughout all the visible Regions of the Universe, in the mutual Assistances, and Returns, which one Globe affords the other, even at the greatest Distance' (Derham 1731, 192). Such vague allusions to large scale mutual influences, however, were increasingly subservient to the teleology of inhabitation, and the 'plurality of worlds' philosophy began to cement itself as the dominant paradigm of astronomical cosmology.

This was becoming the case in the universities as well. The statutes of the Savilian Chair of Astronomy at Oxford, established in 1619, prevented its holders from practicing astrology. In 1718, the then-chair John Keill (1671–1721) published a set of lectures he gave on astronomy with the title *Introductio ad veram astronomiam* ('Introduction to the True Astronomy', English translation 1721). In the fourth lecture, Keill framed an 'admirable magnificent Idea or Notion' of the immense size of the universe. This admirable idea is reached by imagining the indefinite space filled by innumerable suns, each with its own planets, the combination of which 'constitute so many particular Worlds or Systems' (Keill 1748, 40). Each sun performs the same office in its system as our sun does in ours: illustrating, warming and cherishing its attending planets. The Plumian Professorship of Astronomy at Cambridge, meanwhile, has a story about its inception which links it to pluralism. The second holder of that chair, Robert Smith (1689–1768), wrote a note on the fly-leaf of his copy of Huygens' *Cosmotheoros* stating the following:

> I have been well informed that Dr Plume, Archdeacon of Rochester, was so pleased with this book, which the celebrated Mr Flamsteed had recommended to him, as to leave by his will £1800 to found the Plumian Professorship of Astronomy and Experimental Philosophy, which I held many years after Mr Cotes's decease (Edleston 1850, lxxiv–lxxv).

There seems to be no independent corroboration of this story, although it was related by Crowe as fact. It is interesting in itself, however, that Smith associated the Plumian chair with Huygens' theories of ET life, while his own works on optical and musical theory carry none of the astrological implications that so intrigued Kepler a century before.

As the eighteenth century progressed, the number of works of astronomy, theology and natural philosophy that engaged with the question of ET life increased exponentially. Pluralism started to permeate all levels of society: from technical treatises to popular philosophy; from religious sermons to poetry and theatre. Astrology, meanwhile, became increasingly cut off from the top-end of that spectrum. The purpose of this chapter has been to suggest that the rise of pluralism and the marginalisation of astrology were linked in at least one important sense, and that is that important advocates of pluralism, such as More, Flamsteed, Fontenelle, Huygens, Ray and Bentley, consciously and deliberately opposed it to astrology. For these and other pluralist philosophers, the primary reason for God's creation of the planets was to act as habitats for living creatures. The role of influence was generalised and localised to the various suns and satellites, or relegated to a secondary function of all celestial bodies. This reassessment was undertaken at a time when astrology was still highly visible in social and print culture, with links to the intellectual elite via medicine and the almanac trade. There is a tendency still to see the decline of astrology and the resurgence of pluralism as inevitable consequences of heliocentrism, mechanism and then Newtonianism. These examples demonstrate, however, that there was at least some degree of conscious effort to employ analogy and teleology in the service of one theory ahead of the other.

References

Almond, Philip. 2006. Adam, pre-Adamites, and extra-terrestrial beings in early modern Europe. *Journal of Religious History* 30: 163–174.

Baker, George. 1822–1841. *History and antiquities of the county of Northampton*, 2 vols. London: John Bowyer Nichols & Son and John Rodwell.

Bentley, Richard. 1836–1838. In *Works*, ed. Alexander Dyce. 3 vols. London: Francis Macpherson.

Borel, Pierre. 1658. *A New Treatise, Proving a Multiplicity of Worlds*. Trans. D. Sashott. London: John Streater.

Boyle, Robert. 1688. *A disquisition about the final causes of natural things: Where in it is inquir'd, whether, and (if at all) with what cautions, a naturalist should admit them?* London: John Taylor.

Brewster, David. 1855. *Memoirs of the life, writings, and discoveries of Sir Isaac Newton*, 2 vols. Edinburgh: T. Constable and Co.

Burke, John. 1835–1838. *A genealogical and heraldic history of the commoners of Great Britain and Ireland*, 4 vols. London: Colburn.

Burnet, Thomas. 1697 [1690]. *The theory of the Earth*, 3rd ed. 4 bks. London: Walter Kettilby.

Burnett, Charles. 1997. *The introduction of Arabic learning into England*. London: British Library.

Butler, John. 1680. *Hagiastrologia, or, the most sacred and divine science of astrology*. London: John Butler.

———. 1697. *The true state of the case of John Butler, B.D., a minister of the true Church of England, in answer to the libel of Martha, his sometimes wife*. London: John Butler.

Capp, Bernard Stuart. 1979. *Astrology and the popular press: English almanacs 1500–1800*. London: Faber.

Cesalpino, Andrea. 1593. *Quaestionum peripateticarum libri V*. Venice: Giunta.

Cheyne, George. 1705. *Philosophical principles of natural religion*. London: George Strahan.

Clucas, Stephen. 1991. Poetic atomism in seventeenth-century England: Henry More, Thomas Traherne and scientific imagination. *Renaissance Studies* 5: 327–340.

Crawford, I.A. 2018. Widening perspectives: The intellectual and social benefits of astrobiology (regardless of whether extraterrestrial life is discovered or not). *International Journal of Astrobiology* 17: 57–60.

Crowe, Michael J. 1986. *The extraterrestrial life debate, 1750–1900: The idea of a plurality of worlds from Kant to Lowell*. Cambridge: Cambridge University Press.

Cudworth, Ralph. 1678. *The true intellectual system of the universe: Wherein all the reason and philosophy of atheism is confuted, and its impossibility demonstrated*. London: Richard Royston.

Curry, Patrick. 1989. *Prophecy and power: Astrology in early modern England*. Princeton: Princeton University Press.

De Callatay, Godefroid, and Liana Saif. 2017. *The astrological and prophetical cycles in the pseudo-Aristotelian Hermetica and in the Ikhwān al-Ṣafā'*. Presented at the 'Bilan et perspectives des études Sur les encyclopédies médiévales: Orient-occident, le ciel, l'homme, le verbe, l'animal' organized by the ARC project speculum Arabicum, Brugelette, 2017. https://dial.uclouvain.be/pr/boreal/object/boreal:182356. Accessed 28 Nov 2017.

Derham, William. 1720 [1713]. *Physico-theology, or, a demonstration of the being and attributes of God from his works of creation: Being the substance of sixteen sermons preached in St. Mary-le-Bow Church, London, at the Honourable Mr. Boyle's lectures, in the years 1711, and 1712: With large notes, and many curious observations*, 5th ed. London: W. and J. Innys.

———. 1731 [1714]. *Astro-theology: Or, a demonstration of the being and attributes of God, from a survey of the heavens*, 6th ed. London: W. Innys.

Dick, Steven J. 1982. *Plurality of worlds: The origins of the extraterrestrial life debate from Democritus to Kant*. Cambridge/New York: Cambridge University Press.

Edleston, J., ed. 1850. *Correspondence of Sir Isaac Newton and Professor Cotes: Including letters of other eminent men*. London: John W. Parker.

Farley, John. 1977. *The spontaneous generation controversy from Descartes to Oparin*. Baltimore: Johns Hopkins University Press.

Fletcher, John E. 1970. Astronomy in the life and correspondence of Athanasius Kircher. *Isis* 61: 52–67.

Fontenelle, Bernard le Bovier de. 1683. *Nouveaux dialogues des morts*, 2 vols. 2nd ed. Lyon: Thomas Amaulry.

———. 1809 [1686]. *Conversations on the plurality of worlds*. London: Printed for Lackington Allen & co.

Foster, Joseph. 1891. *Alumni oxonienses: The members of the University of Oxford, 1500–1714*. Oxford/London: Parker and Co.

Gabbey, Alan. 1990. Henry More and the limits of mechanism. In *Henry More (1614–1687) tercentenary studies*, ed. Sarah Hutton, 19–36. Dordrecht: Kluwer Academic Publishers.

Geneva, Ann. 1995. *Astrology and the seventeenth century mind: William Lilly and the language of the stars*. Manchester: Manchester University Press.

Giglioni, Guido. 2008. The cosmoplastic system of the universe: Ralph Cudworth on Stoic naturalism. *Revue d'histoire des sciences* 61: 313–331.

Glomski, Jacqueline. 2015. Religion, the cosmos, and counter-reformation: Athanasius Kircher's *Itinerarium exstaticum*. In *Acta conventus neo-latini monasteriensis*, ed. Astrid Steiner-Weber and K.A.E. Enenkel, 227–236. Leiden: Brill.

Goodrum, Matthew R. 2002. Atomism, atheism, and the spontaneous generation of human beings: The debate over a natural origin of the first humans in seventeenth-century Britain. *Journal of the History of Ideas* 63: 207–224.

Grew, Nehemiah. 1701. *Cosmologia sacra, or a discourse of the universe as it is the creature and kingdom of God*. London: W. Rogers, S. Smith and B. Walford.

Guerlac, Henry, and M.C. Jacob. 1969. Bentley, Newton, and providence: The Boyle lectures once more. *Journal of the History of Ideas* 30: 307–318.

Harrison, Peter. 2002. Voluntarism and early modern science. *History of Science* 40: 63–89.

———. 2009. Voluntarism and the origins of modern science: A reply to John Henry. *History of Science* 47: 223–231.

———. 2013. Laws of nature in seventeenth-century England: From Cambridge Platonism to Newtonianism. In *The divine order, the human order, and the order of nature: Historical perspectives*, ed. Eric Watkins, 127–148. New York: Oxford University Press.

Hatch, Robert Alan. 2017. Between astrology and Copernicanism: Morin – Gassendi – Boulliau. *Early Science and Medicine* 22: 487–516.

Haugen, Kristine Louise. 2011. *Richard Bentley: Poetry and enlightenment*. Cambridge, MA/ London: Harvard University Press.
Henry, John. 1990. Henry More versus Robert Boyle: The spirit of nature and the nature of providence. In *Henry More (1614–1687) tercentenary studies*, ed. Sarah Hutton, 55–76. Dordrecht: Kluwer Academic Publishers.
———. 2004. Metaphysics and the origins of modern science: Descartes and the importance of laws of nature. *Early Science and Medicine* 9: 73–114.
———. 2009. Voluntarist theology at the origins of modern science: A response to Peter Harrison. *History of Science* 47: 79–113.
Hunter, Michael. 1995. Science and astrology in seventeenth-century England: An unpublished polemic by John Flamsteed. In *Science and the shape of orthodoxy: Intellectual change in late seventeenth-century Britain*, ed. idem, 245–285. Woodbridge: Boydell Press.
Huygens, Christiaan. 1722 [1698]. *The Celestial Worlds Discover'd*. Trans. John Clarke, 2nd ed. London: James Knapton.
Jaki, Stanley L. 1978. *Planets and planetarians: A history of theories of the origin of planetary systems*. Edinburgh: Scottish Academic Press.
Jenkin, Robert. 1700 [1696–1697]. *The reasonableness and certainty of the Christian religion*, 2nd ed. 2 Bks. London: P.B. and R. Wellington.
———. 1721 [1696–97]. *The reasonableness and certainty of the Christian religion*, 5th ed. 2 vols. London: printed by W.B. for Richard Sare.
Keill, John. 1748 [1718]. *An introduction to the true astronomy: Or, astronomical lectures, read in the astronomical school of the University of Oxford*, 4th ed. London: Henry Lintot.
Kircher, Athanasius. 1660 [1656]. *Iter exstaticum coeleste*. Wurzburg: Johann Andrea Endter.
Koyré, Alexandre. 1957. *From the closed world to the infinite universe*. Baltimore: Johns Hopkins Press.
Kubrin, David. 1967. Newton and the cyclical cosmos: Providence and the mechanical philosophy. *Journal of the History of Ideas* 28: 325–346.
Lambert, Johann Heinrich. 1761. *Cosmologische Briefe über die Einrichtung des Weltbaues*. Augsburg: Klett.
———. 1976. *Cosmological Letters on the Arrangement of the World-Edifice*. Trans. Stanley L. Jaki. Edinburgh: Scottish Academic Press.
Lovejoy, Arthur O. 1948 [1936]. *The great chain of being: A study of the history of an idea*. Cambridge, MA: Harvard University Press.
Monod, Paul Kléber. 2013. *Solomon's secret arts: The occult in the age of enlightenment*. New Haven; London: Yale University Press.
More, Henry. 1681. *Tetractys anti-astrologica, or, the four chapters in the Explanation of the Grand Mystery of Holiness: Which contain a brief but solid confutation of judiciary astrology, with annotations upon each chapter: wherein the wondrous weaknesses of John Butler, B.D. his answer called A Vindication of Astrology, & c. are laid open...* London: J.M.
Morin, Jean-Baptiste. 1661. *Astrologia gallica*. The Hague: Adriaan Vlacq.
Parker, Samuel. 1678. *Disputationes de Deo et providentia divina ....* London: J. Martyn.
Poole, William. 2015. Lucretianism and some seventeenth-century theories of human origin. In *Lucretius and the early modern*, ed. David Norbrook, Stephen Harrison, and Philip Hardie. Oxford: Oxford University Press.
Raffaelli, Enrico G. 2001. *L'oroscopo del mondo: il tema di nascita del mondo e delprimo uomo secondo l'astrologia zoroastriana*. Milan: Mimesis.
Raven, Charles E. 1986. *John Ray, naturalist: His life and works*. Cambridge: Cambridge University Press.
Ray, John. 1714 [1691]. *The wisdom of God manifested in the works of the creation*, 6th ed. London: William Innys.
Reid, Jasper William. 2012. *The metaphysics of Henry More*. Dordrecht: Springer.
Ross, Alexander. 1646. *The new planet no planet, or, the Earth no wandring star, except in the wandring heads of Galileans*. London: J. Young.

Rutkin, H. Darrel. 2006. Why Newton rejected astrology: A reconstruction, or "Newton's comets and the transformation of astrology": 20 years later. *Cronos* 9: 85–98.

Saumaise, Claude. 1648. *De annis climactericis et antiqua astrologia diatribae*. Leiden: Elzevir.

Schaffer, Simon. 1987. Newton's comets and the transformation of astrology. In *Astrology, science, and society: Historical essays*, ed. Patrick Curry, 219–243. Woodbridge/Wolfeboro: Boydell Press.

Selden, John. 1661. *Theanthropos: Or, God made man: A tract proving the nativity of our Saviour to be on the 25. of December*. London: F.G.

Sellars, John. 2012. Stoics against stoics in Cudworth's *A treatise of freewill*. *British Journal for the History of Philosophy* 20: 935–952.

Sheppard, Kenneth. 2015. *Anti-atheism in early modern England, 1580–1720: The atheist answered and his error confuted.* Leiden: Brill.

Simonutti, Luisa. 2001. Spinoza and Boyle: Rational religion and natural philosophy. In *Religion, reason and nature in early modern Europe*, ed. Robert Crocker, 117–138. Dordrecht/London: Kluwer Academic.

Sturmy, Daniel. 1711. *A theological theory of a plurality of worlds: Being a critical, philosophical, and practical discourse, concerning visible or material worlds*. London: D. Brown and J. Walthoe.

Thomas, Keith. 1971. *Religion and the decline of magic: Studies in popular beliefs in sixteenth and seventeenth century England.* London: Weidenfeld & Nicolson.

Vermij, Rienk. 2016. Seventeenth-century Dutch natural philosophers on celestial influence. In *Unifying heaven and earth: Essays in the history of early modern cosmology*, ed. Miguel Á. Granada, Dario Tessicini, and Patrick J. Boner, 291–315. Barcelona: Edicions Universitat Barcelona.

Vermij, Rienk, and Hiro Hirai. 2017. The marginalization of astrology: Introduction. *Early Science and Medicine* 22: 405–409.

Westman, Robert S. 2011. *The Copernican question: Prognostication, skepticism, and celestial order*. Berkeley/London: University of California Press.

Wood, Anthony. 1691–1692. *Athenae Oxonienses*, 2 vols. London: Thomas Bennet.

Wordsworth, C., ed. 1842. *The correspondence of Richard Bentley*. London: John Murray.

Wright, Peter. 1975. Astrology and science in seventeenth-century England. *Social Studies of Science* 5: 399–422.

Chapter 7
Conclusion

Thus, have we found a true astrology;
TThus, have we found a new, and noble sense,
IIn which alone stars govern human fates.

Edward Young, *Night Thoughts* (Young 1813, I, 305)

Abstract The concluding chapter begins with a line from Edward Young's poem *Night Thoughts*: 'Thus, have we found a true astrology/Thus, have we found a new, and noble sense/In which alone stars govern human fates.' After summarising the book's key findings, restating the two main theses, and suggesting areas for further research, the chapter turns to the topic of astrobiology. The question is posed: 'Is Astrobiology a New Astrology?' Arguments are made that a belief or interest in extraterrestrial life replaced astrology in several important dimensions. The conclusion then debates whether it is useful to think of this in terms of a Kuhnian shift from a paradigm of celestial influence to one of celestial inhabitation, and what such a theory might entail for the histories of astrology and the extraterrestrial life debate.

Keywords Astrology · Extraterrestrial life · Plurality of worlds · Astrobiology · Anthropology · Paradigm shift

7.1 Summary

This book has aimed to present a conjoined and comparative history of astrology and pluralism in the early modern era. The approach taken has been to isolate the two fundamental ideas—celestial influence and celestial inhabitation—and trace the relationship and interactions between them in the works of individual natural

© Springer Nature Switzerland AG 2019

J. E. Christie, *From Influence to Inhabitation*, International Archives of the History of Ideas Archives internationales d'histoire des idées 228,
https://doi.org/10.1007/978-3-030-22169-0_7

philosophers throughout this period. The selection of individuals discussed has not been random, nor has it been completely representative. The varyingly detailed case studies in this volume were chosen in order to fill historical gaps, to challenge or complement certain aspects of the established narrative and interpretation, and especially to illustrate connections between the two histories. Some of these connections were recurrent enough to suggest trends, and two of these trends form the basis for the historical arguments advanced by this book. The first is that evolving theories of celestial influence formed part of the catalyst for thinking about ET life; the second is that by the latter half of the seventeenth century, anti-astrological thinkers were using pluralism to negate the teleological claims of the astrologers.

We began with Plutarch's *De facie*, and from there proceeded to Nicholas of Cusa's *De docta ignorantia*, via a discussion of the functions of celestial influence within the largely Aristotelian cosmology of the medieval and Renaissance periods. Plutarch and Cusa were the two most important pre-Copernican sources for ideas about life on other celestial bodies within our cosmos. In both cases celestial influence was involved. In Plutarch, this took the form of Stoic ideas about celestial bodies nourishing one another, and the suggestion that the earth may have a reciprocal influence on the moon which helps to support life. In Cusa, it took the form of a question: if the influences of the stars produce living creatures in the regions of the earth, and the stars receive influences from each other, why shouldn't living creatures be produced in their regions as well? Both of these ideas suggest a universality or at least a commonality to the operations of nature, long before this would become a central tenet of the mechanical philosophies.

Cusa was exceptional in this regard, as, indeed, he was in many others. The Aristotelian/Ptolemaic cosmos instituted a strict ontological divide between the sublunary and superlunary regions, limiting the process of generation and corruption, and thus life as we now understand it, to the terrestrial, elemental sphere. Within this cosmos, the function of the heavenly bodies was understood in terms of their influence on the earth and therefore their usefulness to humanity. They constituted a celestial causal chain linking the central earth to the outermost *primum mobile* and empyrean heaven. The particular mechanisms of this causal chain in regard to biological processes were developed by Alexander of Aphrodisias and the Arabic philosophers into an astrologised version of Aristotle's theories of generation and corruption, and from the Arabic sources and commentaries it was absorbed into Latin Christendom. Thus, the 'science of the stars' in medieval and Renaissance natural philosophy involved the study of both the motions of the celestial bodies as well as their effects. The Lutheran educational reformer Philip Melanchthon is an example of one particularly influential figure who saw the study of astrology as a religious endeavour; a way of understanding God's providence through the study of his celestial instruments.

The threat to astrology posed by Copernicus' heliocentric theory was not as evident to people at the time as it was to Henry More and Richard Bentley in the following century. Figures like Rheticus, Philip van Lansberge and others understood and explained Copernicanism in astrological terms. The increased size of heliocentric systems, however, did lead some to introduce a teleology of inhabitation, at least in terms of spiritual or immaterial beings. Tycho Brahe's published accounts of comets and a new star suggested that the heavens were not incorruptible, and so-called *novatores* like Patrizi started to develop alternatives to Aristotle's physics, abandoning the four elements and bridging the gap between the terrestrial and celestial worlds. Patrizi was critical of judicial astrology, but believed that stars and planets, as corporeal bodies, must have some active virtues. His endorsement of the 'Pythagorean' theory that each star constitutes a world arose out of his conviction that these bodies must act within and amongst themselves with the purpose of creating and sustaining living (albeit *ethereal*) creatures. The mathematician and astrologer Benedetti took a different approach, but his theory about ET life was still tied to ideas about celestial influence. His interpretation of Copernicanism led him to suggest that the planets may be moons orbiting around other unreflecting bodies like the earth. These planets should therefore be understood as mirrors, directing sunlight onto their respective 'earths'.

An entire chapter was dedicated to the terrestrial and celestial physics of William Gilbert, in part because the level of appreciation and understanding of his philosophy does not match the level of its importance to the histories of astrology and pluralism. The astrological nature and implications of magnetism are very apparent in Gilbert's own works, but his theory would later act as a stepping stone for the assimilation of occult action at a distance into a de-astrologised physics. The belief that the planetary forms, of which magnetism was just one example, were both influential and motive was crucial to the logic of Gilbert's world-system. Their various influences contributed to the diversification of the material and hence the generation of living creatures on the surfaces of each of the globes, and the motion of these globes could be understood as the attempt to constantly maintain an ideal mixture of influences.

This combination of influence and ET life as a way of understanding celestial phenomena was replicated by Kepler. His analysis of the astrological import of the nova of 1604 led him to suggest that this celestial sign may be intended for the inhabitants of a completely different globe. His combination of harmonic theory with astronomy and astrology, meanwhile, led him to speculate about beings on the sun who alone could appreciate the harmonic ratios produced by the orbital velocities of its surrounding planets. Here again is an instance of influence and inhabitation combining as an explanation for celestial motion. In both cases, moreover, the action of celestial influence at places other than the earth was evidence for Kepler that those places should be inhabited. The importance given to Kepler in this book is a result largely of the prominent place he holds in the histories both of astrology and pluralism. The productive connection between these ideas in his philosophy will hopefully be useful, therefore, in demonstrating the potential benefit of considering these histories together.

In Chap. 5, we reverted back to a broader survey of mid-seventeenth-century cosmologies with an eye on the symbiotic relationship between theories of influence and theories of inhabitation. By this point, thanks largely to the telescopic observations of Galileo and the work on lunar inhabitation by Wilkins, pluralism was founded upon more general 'Copernican' arguments, based on the analogy of the earth to the other planets and the difficulty of explaining the newly observed, far-distant celestial objects in terms of human utility. Still, varying conceptions of celestial influences, not limited to those of the sun and the moon, remained in these philosophies. For Borel, Otto von Guericke, Le Grand, Gadroys and Wittie, the reciprocal exchange of influences among the celestial bodies preserved a sense of harmonious cosmic unity in a system that risked becoming fragmented. There seemed to be a reluctance as well to conclude that distant celestial bodies had *absolutely* no relation to the earth. The desire for humanity to be connected to the larger cosmos, it seems, could not easily be brushed aside.

The limited and, again, selective survey of seventeenth-century natural philosophies has hopefully demonstrated that there was nothing incompatible about a belief in astrological influence and a belief in ET life. The attempts of Kepler and Gadroys to construct functioning systems of astrological theory within systems otherwise inclined to pluralism proves as much. The more salient point for the history of pluralism, however, is that the (re-)emergence of this tradition took place at a time when the astrological paradigm, if we want to call it that, was still dominant, and so its growth should be understood as taking place within and alongside discussions of celestial influence. Sherburne's translation of Manilius is a case in point. The appendices, which include references to pluralist and astrological theories taken from contemporary sources, demonstrate that the readership would have been interested in both subjects, and probably would not have perceived any ancient/modern divide between them. The discussion of ET life and astrology by the same people and in the same contexts provides support for the argument that it was opponents, both of pluralism and of astrology, who emphasised the dichotomy between the two, creating in the process the view that each was logically tied to its own philosophical system.

We see this in the attacks by White, More, Flamsteed and Bentley, when they targeted the astrologers' claim that the stars would have been created in vain if they didn't exert significant influences upon the earth. Instead of relying only on other, non-astrological benefits for humanity, they all declared it arrogant to think that the entire universe has been created for us. They claimed that astrology relied either on an idolatrous belief in god-like planets, or a geocentric cosmology, or both; the belief in ET life, on the other hand, was associated with the Copernican system, the discoveries of the telescope, and a more mechanical understanding of the workings of nature. On the other side of the debate, figures such as Ross, Morin, Butler and Jenkin, in defending astrology or criticising pluralism, had to reassert an anthropocentric teleology—an approach that was less and less convincing to a late-seventeenth-century readership. This antagonism began to erode the symbiosis of astrological and pluralist ideas.

In the Newtonian cosmology of Bentley, this symbiosis was abandoned completely. The mutual exchange of celestial influence—a natural means by which celestial bodies maintain the stable yet varied conditions best suited to habitation—was too naturalistic, relying too heavily on secondary causes and an overly-mechanical conception of life. Unlike Digby, whose attempt to explain biological generation mechanically accentuated the uniformity of nature, Bentley's emphasis on the immediate and arbitrary action of God shifted focus back to plenitude and teleology. Our discussion of the period finished with a review of Newtonian natural theology, looking not so much at the details of speculation about ET life, which has been covered more by other historians, as at the absence of astrology. Comparing this project with earlier syntheses of natural philosophy and theology, like those of Thomas Aquinas and Melanchthon, makes clear just how dramatic the teleological shift from influence to inhabitation was.

The historical narrative presented in this book has something of an Oedipal theme. The marriage of astronomy to astrology contributed to the conception of pluralism. Pluralism then developed, and in doing so it facilitated the decline of astrology, eventually taking its place at astronomy's side. It would be a mistake, however, to think of this as some sort of grand or self-standing narrative. The strands of natural philosophy reconstructed here are currents and eddies within the larger stream of the history of cosmology. The demarcation of these strands is intended to contribute to, rather than rewrite, the histories of astrology and the ET life debate in this period. For example, the close association of ideas about celestial influence—and the astrological tradition more broadly—with questions related to celestial physics and inhabitation is an important contextual and intellectual accompaniment to the histories of Dick and Crowe, as well as more recent article and monograph contributions that have been spurred by the current resurgence of interest in exoplanets and alien life.[1] As for the history of astrology, the active and passive role played by cosmological pluralism should be counted alongside other philosophical, theological, political, social and institutional factors that contributed to its marginalisation.[2]

7.2 Further Questions

On top of these somewhat modest objectives, there are other issues that a combined history of astrology and pluralism can help to elucidate. One of these issues involves the different attempts to develop a rationale for celestial measurements and motion within a non-Aristotelian/Ptolemaic cosmos. Earlier theories of solid celestial spheres or individual planetary intelligences were predicated on a teleology of influence and an ontological terrestrial/celestial divide: a hierarchical nesting of perfect bodies moving in perfect circular motion, acting via heat and light through a celestial medium as efficient and/or specific causes on the imperfect terrestrial world. In the philosophies of Bruno, Gilbert, Hill, von Guericke and, to certain extent, Hobbes, there was a resurgence of Stoic and Platonic concepts, preserved in a pluralist context by Plutarch, which depicted the celestial bodies as living creatures that move according to an intrinsic and biological inclination towards preservation and

[1] See, for instance, some of the articles on history published in the leading journals of astrobiology, such as Brake 2006; Briot 2013.

[2] Discussions on these separate factors can be found in the various contributions in Dooley 2014.

betterment. This combination of influence and inhabitation resulted in what Paul-Henri Michel called, in the case of Bruno, an astrobiology—a celestial ecosystem in which the spacings and motion of the primary globes could be understood as the result of those bodies mitigating and capitalising on environmental factors. This understanding was then overturned by Cartesian vortices, which explained the motion of the celestial bodies and the distances between them mechanically, and Newtonian gravity, which explained motion according to a physical force law, and distances according to a notion of divine providence premised on the idea of God's absolute will.

This book did not spend much time on the particular nature of lunar or planetary inhabitants. This is partly because most pluralist writers in this period avoided speculating about the specific constitutions or appearances of ET beings, and partly because instances where they did, like in the fictions of Godwin, Cyrano and Kepler, have been well-studied.[3] We did, however, encounter certain writers wondering about the general nature of ET life, specifically in regard to the question of rationality. As Dick observed, most of the interest in the seventeenth and eighteenth centuries concerned intelligent life, with more general biological debate arising in later periods (Dick 1982, 189). In the traditional understanding, however, humanity was an *infima species*, holding exclusive rights to the 'rational animal' branch of the Porphyrian tree. The potential theological problems arising from the possibility of ET life brought this question to the fore. Some thought that the planets may house beings that are superior to mankind; beings which fill up the great chasm in the chain of being between men and God. Others proposed that the planets may serve as settings for the different temporal stages of humanity's soteriological journey. How astrology and pluralism relate to one another in terms of the temporal dimension of cosmology may be a promising area for future research.

What we focused on here, however, were the theories of White and Bentley that widened out humanity's step on the *scala naturae*, suggesting that man is just one species of rational animal, and that the planets are populated by others, distinguished by different combinations of rational soul and corporeal body. The growing engagement with the plurality of worlds philosophy led to dramatic developments in the definition of humanity and its place in the universe.[4] It may also have impacted, or been influenced by, ideas about mankind's relationship to the other forms of life on the earth itself.[5] The question of whether the universe had been created to serve humanity involved those things 'above' mankind in the cosmos, as well as those things 'below' him. These discussions were taking place, moreover, at a time when European civilisation was discovering new areas of the globe, complete with people, animals and plant life different and yet similar to those closer to home.[6] As

[3] It has also been argued that maintaining a sceptical or agnostic position regarding the specifics of ET life allowed these authors to adjust to the arguments of their critics. Matytsin 2013.

[4] For more on this particular issue, see Brooke 2000; Dick 2006; Farman 2012; Packer 2015; Christie 2018.

[5] See Wolloch 2000, 2002.

[6] See in particular Campbell 1999.

knowledge of the terrestrial sphere increased and blank areas on the map were filled in, the heavens, which had seemed fully charted before the telescope, revealed further frontiers—locations that may be just as replete with life, albeit for now beyond the scope of observation.

There were avenues for conjecture, however, and one of them related to astrology. It is not particularly surprising that early ideas about the nature of living beings on the other planets did not depart dramatically from the mythological and astrological associations of those planets. Fontenelle's amorous Venusians and phlegmatic Saturnites vary only a little from the *planetenkinder* of medieval and Renaissance astrology (Fontenelle 1809, 78, 98). These associations lasted a long time, with the war-like nature of Martians being the most recurring and well-known example. In this book, we have noticed that there were in fact more logical motives to these speculations. Cusa suggested that solar beings were intellectual and fiery while lunar beings were watery and aerial, basing his belief on the influence of each of those bodies. The assumption was that one could make surmises about the nature of a celestial body based on the influence that that body emits. Plutarch had done something similar, arguing that there must be moisture present on the moon because the earth receives a moistening influence from it. Whereas astrologers, like Kepler, had once seen a planet's colour as a clue to the quality of its influence, pluralist investigators saw it as an indication of material constitution, and drew inferences about its possible inhabitants. Exoplanet spectroscopy aims to do the same today.

This brings us back to the question of the history of astrobiology. The subtitle of this book refers to a transformation of astrobiology in the early modern period. Most historians of astrobiology, of which there is a small but increasing number, would consider that astrobiology only emerged in this period. This is because, at the moment, the history of astrobiology is effectively the history of exobiology—that is, the history has not expanded to match the broader scope of the discipline. Astrobiology is the study of life on earth as well as ET life, with some advocates giving it the broader and more poetic objective of seeking to understand life in a cosmic context. This ambition is the common thread that unites all the individuals and texts examined in this thesis. Astrology was the pre-modern astrobiology, if we care to think of it like that, uniting astronomy, biology and anthropology in order to understand life on this earth in relation to its celestial setting. It was over the course of the long-seventeenth century that astrobiology transformed, turning away from the question of influence towards that of inhabitation.

7.3 Is Astrobiology a New Astrology?

This raises an interesting and yet difficult question. Taking as a starting point those lines from Edward Young's *Night Thoughts* (1745), which describes the knowledge of an immense universe full of inhabited worlds as 'a new Astrology', we might ask to what extent, or in what ways, a belief in ET life replaced a belief in astrology from the eighteenth century onwards. The main concern of this book has been teleology. It has hopefully been demonstrated by now that the answer to the question 'what are the celestial bodies for?' changed dramatically in this period. The standard answer before the seventeenth century was that they were created to govern the earth via their influences, while the standard answer from the eighteenth century onwards was that they were created to be habitats for other living creatures. There are, however, several important caveats to be made to this generalisation. One is that influence was not always their sole purpose, even within a Christian/Aristotelian cosmos, and there were always critics of astrology who tried to downplay the influences they might have. By the same token, inhabitation was not the only teleological principle within Newtonian cosmology. There is the issue of primary and secondary uses, for example, and instances of celestial bodies and phenomena, like comets and of course the suns themselves, which existed purely to influence other bodies. It is perhaps better, then, to think of the balance of power shifting among multiple teleologies, rather than the hegemony of any particular one.

It is fair to say, however, that inhabitation began to take over a conversation that was once dominated by influence. Another way we might think that pluralism replaced astrology is in its role as the conjectural or speculative side of astronomy. Huygens is an example of one professional astronomer who largely ignored astrology and then devoted a whole work to conjectures about planetary inhabitants. The research of Crowe on the eighteenth and nineteenth centuries has showed that many of the most famous figures in the history of astronomy were deeply concerned with, and wrote extensively on, pluralist questions. Moreover, he has argued that many such astronomers were first attracted to their field of study by pluralist writings, and that the motivation behind many examples of astronomical patronage and instrument building was the desire to investigate questions about ET life. 'The attractiveness of some pluralist theories', suggests Crowe, 'may be attributed to their unfalsifiability, flexibility, and richness in explanatory power' (Crowe 1986, 548). This statement could easily be made about astrological theories as well. Perhaps a helpful tool in this regard would be to think of them both as examples of *themata*, as Gerard Holton defines them: examples of 'non-scientific' commitments which underlie scientific practice.[7]

There are several areas where a belief in ET life did not and could not replace astrology, and the most obvious one concerns utility. We have largely limited ourselves to the subject of astrology's natural philosophical foundations, i.e. celestial influence, but astrology proper was a predictive art, and it was in this practical dimension that astrology permeated and influenced such large swathes of society and culture. While speculations about life on the other planets might have intrigued a new generation of astronomers, it in no way encroached upon astrology's practical territory. As Anselm lamented in Fontenelle's dialogue, people will never be dis-

[7] Holton gives several examples of now discredited *themata*. See Holton 1988, 14: 'ideas such as macrocosmic-microcosmic correspondence, inherent principles, teleological drives, action at a distance, space-filling media, organismic interpretations, hidden mechanisms, or absolutes of time, space, and simultaneity'. Celestial influence and inhabitation are both implicated here, although only one is still of important scientific interest.

abused of things that regard the future, and so the astrologer's trade will always hold good (Fontenelle 1683, I, 258).

That is not to say that pluralism had no popular appeal. We have already mentioned the success of Fontenelle's *Entretiens*, but many other pluralist works, both fiction and non-fiction, had a wide readership. What is quite interesting is that pluralist ideas started to seep into the mainstream at a time when popular astrology and the almanac trade, in England at least, were still near their peak. Anna Marie Roos' analysis of English works of popular science and subscription newspapers, such as the *Athenian Mercury*, reveals an interest both in astrological questions and in the possibility of a world in the moon (Roos 2001, 244–45). The general trend that she depicts, however, is one of 'desacralization', with astrology being increasingly socially discredited. Yet theories of ET life were not above criticism in the public sphere either. Indeed, as astrologers painted pluralism as a modern conceit, and advocates of the new astronomy derided astrology as irreligious superstition, it seems that both at times were scoffed at by a public who didn't see as strict a divide between the astronomer and the astrologer as we do today.[8] Pluralism, however, would eventually begin to take over as an important link between the new astronomical science and the wider, increasingly literate population. Indeed, Crowe has argued that embellishing texts with talk of ET life became a sure-fire way for authors to interest the public in astronomy (Crowe 1986, 556). This state of affairs continues today. Astronomical discoveries are more likely to feature in the news if there is some connection to the search for ET life, and such articles are usually rife with exaggeration and hyperbole.

The other major area of popular engagement is of course science fiction. The lunar fictions of Kepler, Godwin and Cyrano are either foundational texts or precursors of this genre, depending on the flexibility of your definition.[9] More examples of extraordinary voyages followed in the eighteenth and nineteenth centuries, before Jules Verne and H. G. Wells took the genre to new heights, leading up to the 'Golden Age' of sci-fi in the mid-twentieth century. ET characters are in most cases a sufficient, if not necessary, condition for inclusion in the sci-fi genre, and here again we might draw a link back to astrology. Astrological types usually involved the isolation and exaggeration of a human personality trait, and it is in this guise of psychological archetypes that astrology was utilised in medieval and Renaissance literature.[10] The figure of the alien often plays a similar role in sci-fi, being a physical embodiment of exaggerated aspects of the human.[11] In the television series 'Star

[8] See, for instance, Horne 1983. On the increasing satirisation of astrologers, see Reeves 2014.

[9] Carl Sagan and Isaac Asimov reputedly considered Kepler's *Somnium* the first work of science fiction, a title which has stuck. See Christianson 1976; Bozzetto and Evans 1990. All these works are usually discussed in the opening chapters of histories of the genre, such as Aldiss 1973; Suvin 2016 [1979]; Roberts 2006. Other works push the genre's timeline even further back. See Rogers and Stevens 2015; Kears and Paz 2016.

[10] See, for example, Wood 1970; Eade 1984.

[11] See Malmgren 1993.

Trek', created by Gene Roddenberry in the 1960s, the Vulcan and Klingon species, for example, are embodiments of logic and violence respectively.

In this setting, the value of humanity lies in the fact that it embodies a balance of these extremes. A human is more passionate than a Vulcan and more compassionate than a Klingon. As we noticed in Fontenelle and Huygens, humanity went from being the centre of the universe to being its mean or median average. This idea was further developed in an early work by Immanuel Kant (1724–1804). The third part of his *Allgemeine Naturgeschichte und Theorie des Himmels* ('General History of Nature and Theory of the Heavens', 1755) contained an attempt to make reasonable conjectures about the nature of the inhabitants of the other planets in the solar system.[12] He argued that intellectual capacities were dependent upon the relative density and coarseness of the fluids and fibres of the cerebral nerves. Seeing as how material increases in density as you approach the centre of gravity in the solar system, Kant concluded that the inhabitants of the outer planets would be much more intelligent and perfect than those on the inner planets (Kant 1981, 189). Humanity holds the middle rung in this particular scale of being, and Kant believed we should be grateful for it. Our view of the universe encourages both pride and humility, depending on which way you look (Kant 1981, 190).

Kant's discussion of planetary inhabitants creates an intelligible order to our immediate celestial environment and helps us to understand our place within it. Again, we can think of this as fulfilling a function once filled by astrological ideas. In that case, the universe was thought to increase in perfection as it receded from a terrestrial centre of gravity, but at the same time the celestial hierarchy formed a causal chain proceeding from the highest heaven down to the earth. When humanity looked up it was inspired to humility and awe in the face of such perfect creations, but was also grateful for the care taken in the regulation of sublunary affairs. Celestial influence and celestial inhabitation are both modes of thinking that are employed to understand human nature and human concerns within a cosmic context. Even more specifically, they are both attempts to comprehend the chain of being. This chain not only depicted a succession of life-forms from the lowest to the most God-like, it also represented a hierarchy of forces that the knowledgeable practitioner could manipulate. With the shift from influence to inhabitation, the chain of being largely lost this practical dimension, but the desire to see it filled in did not diminish.

Within the works of individual thinkers, this book has treated influence and inhabitation as unit-ideas, reconstructing a history of ideas from the pre-Copernican to the post-Newtonian age. Now that we come to a broader comparison of these epochs, a more relevant historiographical foil is Kuhn's theory of scientific revolution and paradigm shift. In a strict understanding of Kuhn's usage, pluralism would probably be considered an offshoot of the Copernican paradigm, rather than a paradigm in itself. But it did play an important role in the building of heliocentric theories into complete cosmologies or conceptual schemes. Kuhn discussed the requirement that a cosmology supply both an explanation of observed phenomena

[12] See Dick 1982, 165–74; Crowe 1986, 47–56; Losch 2016.

and a 'psychologically satisfying world-view' (Kuhn 1957, 7). We might draw a link here to Guthke's emphasis on myth. Likewise Fernand Hallyn's work on poetics in cosmology, and in particular his use of Ricoeur's analogy between the role of myth and the role of heuristic fictions in science. 'A new hypothesis', Hallyn states, 'as long as it is not sufficiently validated and accepted, is formally similar to a *mythos*, to the intrigue or plot of a tragedy' (Hallyn 1990, 13–14). Even today the existence of ET life has not been proven, but, like astrology, it is flexible, difficult if not impossible to falsify, and it can provide intrigue to cosmology and scientific research.

Taking a wide view over the larger structures of the history of cosmology, the idea of a paradigm shift from influence to inhabitation looks similarly intriguing and richly explanatory. There is still need for some reservation, however. This book has focused largely on the period of flux, when both these ideas were up for debate. The claim that either of these was widely accepted and deeply ingrained enough in its own period to be considered a paradigm relies on the work of other historians in each field. The problem is that historians of astrology and historians of pluralism in the twentieth century may have been liable to overstate the case for the importance of their field in the face of disdain or disregard from mainstream history of science. Dick, writing in 1982, bemoaned the fact that the ET life debate was usually viewed with 'more disdain than objectivity, more amusement than serious scholarship' (Dick 1982, 176). This might explain the use of particularly grand and hyperbolic statements, e.g. that 'the assertion of extraterrestrial intelligence may be considered a metaphysical completion of the scientific revolution' (Dick 1982, 188). The history of astrology has been making its case for even longer, and yet still there is the lingering feeling in that camp that the importance of astrology in medieval and Renaissance society is under-studied and under-appreciated.[13] The entrenched self-image of these histories as correctives to a traditional historical narrative might caution us to think that the truth lies somewhere in the middle. This thesis, by inviting a comparison between the two, is vulnerable to attack from both sides.

Nevertheless, when someone with such a comprehensive knowledge of the subject as Lynn Thorndike describes astrology as a widely-held, universal natural law, we are inclined to grant it credence. As for the dominance of pluralism in the post-Newtonian age, we can look to the testimony of the nineteenth-century scientist and historian of science William Whewell (1794–1866). When Whewell wrote an essay arguing against the belief that the planets were inhabited, he seemed well aware of the way that the cosmological landscape had changed over the preceding centuries. 'It will be a curious, but not very wonderful event', he wrote, 'if it should now be deemed as blameable to doubt the existence of inhabitants of the Planets and Stars, as, three centuries ago, it was held heretical to teach that doctrine' (Whewell 1853,

[13]As just one example, note this line from Vermij and Hirai 2017, 406: 'Although the history of astrology has become a blossoming field of research in recent decades, its researchers still have to fight against the prejudice that their subject is a mere superstition or chimaera and, therefore, unworthy of serious consideration.'

iii).[14] The example of Whewell presents an opportunity to look once more at how theological approaches to cosmology shifted over long periods of time. We have discussed how the astrological vision of Manilius presented an ordered, providential system in opposition to the chaotic pluralism of Lucretian atomism. This position was maintained in the medieval and Renaissance periods, with Epicurus often represented as the most dangerous example of philosophical atheism. In the seventeenth century, however, More and Bentley presented pluralism as a theological antidote to the perceived deterministic naturalism and irreligion of astrology.

In the context of nineteenth-century natural theology, the debate over ET life between Whewell and David Brewster saw earlier Newtonian positions reversed.[15] Whewell took a scientific and theological position against the plurality of worlds philosophy in the face of its increasing naturalism in the context of the nebular hypothesis developed by Pierre-Simon Laplace (1749–1827). By this point a belief in ET life had again become associated with an anti-religious, or at least an anti-scriptural, position, in part due to the writings of Jérôme Lalande (1732–1807) and Thomas Paine (1737–1809) (Crowe 1986, 77–80). Darwin's theory of evolution would exacerbate this further, meaning that it was possible not only to explain the formation of the solar system in completely mechanical terms, but also to understand the evolution (if not the ultimate origin) of life in naturalistic terms.

It is a curious event (to borrow his words) that Whewell, an opponent of interest in ET life, was primarily responsible for the demarcation of the scientific disciplines, creating an institutional divide between astronomy and biology that astrobiology has only recently tried to bridge. The modern discipline of astrobiology is little more than 20 years old, but hopefully this anachronistic 'long history' of astrobiology and its transformation in the early modern period can help us to understand its current popularity on a historical, sociological and anthropological level. Astronomers and biologists no longer approach their specimens as bodies created by God for a particular purpose, but by prioritising some lines of enquiry above others, by making implicit judgments about what is interesting or worthwhile, we impose our own teleology on the world around us. In astrobiology, planets and moons are 'for' inhabitation just as they were in natural theology. In her ethnographic work within this discipline, the anthropologist Lisa Messeri talked about the transformation of planets into 'places'—an act which facilitates a mode of connection and offers the possibility to know one's own place in the universe (Messeri 2016, 192).

Astrology is often understood as a semiotic practice, transforming the night sky into a canvas painted with signs and meanings. Messeri describes how, at a conference held at MIT in 2011 on 'The Next 40 Years of Exoplanets', the astrobiologist Sara Seager showed a photo of two children pointing up at a twilight sky, with the following caption: 'We want to show our children, grandchildren, nieces, and nephews a dark sky, point to a star visible to the naked eye, and tell them, "that star has

[14] On Whewell's complex relationship with pluralism, see the chapter in Crowe 1986, 265–99.

[15] See Brooke 1977; Crowe 1986, 300–353.

a planet like Earth"' (Messeri 2016, 1).[16] This demonstrates the desire to be able to look at the night sky and anchor it with new points of meaning, instead of relying on old stellar constellations, long acknowledged to be figments of the human imagination. Astrobiology is in many ways an attempt to turn our modern astronomical cosmography into a true cosmology, using inhabitation as a determiner to superimpose order and meaning onto the fabric of outer space.

This comparison of modern astrobiology with astrology is not meant as a criticism. It is intended to demonstrate that, in an instructive sense, our interest in ET life can be thought of as a modern incarnation of the astrobiological imagination. This functionalist approach, although it has obvious limits, may help to mitigate the seeming antiquarianism of the history of astrology as well as the presentism of the history of the ET life debate. Rather than thinking in normative terms of disenchantment or Enlightenment, we might instead view the decline of astrology and the rise of pluralism as a transformation of astrobiology—a shift from celestial influence to celestial inhabitation as the dominant paradigm for the scientific and poetic approach to an understanding of life in a cosmic context.

References

Aldiss, Brian Wilson. 1973. *Billion year spree: The history of science fiction*. London: Weidenfeld and Nicolson.

Bozzetto, Roger, and Arthur B. Evans. 1990. Kepler's "Somnium"; or, science fiction's missing link ("Le Songe" de Kepler, ou le chaînon manquant de la science fiction). *Science Fiction Studies* 17: 370–382.

Brake, Mark. 2006. On the plurality of inhabited worlds: A brief history of extraterrestrialism. *International Journal of Astrobiology* 5: 99–107.

Briot, Danielle. 2013. Elements for the history of a long quest: Search for life in the universe. *International Journal of Astrobiology* 12: 254–258.

Brooke, John Hedley. 1977. Natural theology and the plurality of worlds: Observations on the Brewster--Whewell debate. *Annals of Science* 34: 221–286.

———. 2000. "Wise men nowadays think otherwise": John Ray, natural theology and the meanings of anthropocentrism. *Notes and Records of the Royal Society of London* 54: 199–213.

Campbell, Mary B. 1999. *Wonder & science: Imagining worlds in early modern Europe*. Ithaca/London: Cornell University Press.

Christianson, Gale E. 1976. Kepler's *Somnium*: Science fiction and the renaissance scientist. *Science Fiction Studies* 3: 79–90.

Christie, James E. 2018. Stepping sideways on the *scala naturae*: Confronting the extraterrestrial in early modern literature. In *Cultural encounters: Cross-disciplinary studies from the late middle ages to the Enlightenment*, ed. Désirée Cappa et al., 127–144. Wilmington: Vernon Press.

Crowe, Michael J. 1986. *The extraterrestrial life debate, 1750–1900: The idea of a plurality of worlds from Kant to Lowell*. Cambridge: Cambridge University Press.

Dick, Steven J. 1982. *Plurality of worlds: The origins of the extraterrestrial life debate from Democritus to Kant*. Cambridge/New York: Cambridge University Press.

[16]On the tension between an ostensibly anti-teleological science and the need for science popularisers to appeal to meaning and values, see Helsing 2016.

———. 2006. Anthropology and the search for extraterrestrial intelligence: An historical view. *Anthropology Today* 22: 3–7.
Dooley, Brendan Maurice, ed. 2014. *A companion to astrology in the Renaissance*. Leiden: Brill.
Eade, J.C. 1984. *The forgotten sky: A guide to astrology in English literature*. Oxford: Clarendon Press.
Farman, Abou. 2012. Re-enchantment cosmologies: Mastery and obsolescence in an intelligent universe. *Anthropological Quarterly* 85: 1069–1088.
Fontenelle, Bernard le Bovier de. 1683. *Nouveaux dialogues des morts*, 2 vols. 2nd ed. Lyon: Thomas Amaulry.
———. 1809 [1686]. *Conversations on the plurality of worlds*. London: Printed for Lackington Allen & co.
Hallyn, Fernand. 1990. *The poetic structure of the world: Copernicus and Kepler*. New York/ Cambridge, MA: Zone Books/Distributed by MIT Press.
Helsing, Daniel. 2016. Uses of wonder in popular science: *Cosmos: A personal voyage* and the origin of life. *International Journal of Astrobiology* 15: 271–276.
Holton, Gerald. 1988 [1973]. *Thematic origins of scientific thought: Kepler to Einstein*. Revised ed. Cambridge, MA/London: Harvard University Press.
Horne, William C. 1983. Curiosity and ridicule in Samuel Butler's satire on science. *Restoration: Studies in English Literary Culture, 1660–1700* 7: 8–18.
Kant, Immanuel. 1981. *Universal Natural History and Theory of the Heavens*. Trans. Stanley L. Jaki. Edinburgh: Scottish Academic Press.
Kears, Carl, and James Paz, eds. 2016. *Medieval science fiction*. London: King's College.
Kuhn, Thomas S. 1957. *The Copernican revolution: Planetary astronomy in the development of Western thought*. Cambridge, MA/London: Harvard University Press.
Losch, Andreas. 2016. Kant's wager: Kant's strong belief in extra-terrestrial life, the history of this question and its challenge for theology today. *International Journal of Astrobiology* 15: 261–270.
Malmgren, Carl D. 1993. Self and other in SF: Alien encounters. *Science Fiction Studies* 20: 15–33.
Matytsin, Anton. 2013. Scepticism and certainty in seventeenth- and eighteenth-century speculations about the plurality of worlds. *Science et Esprit* 65: 359–372.
Messeri, Lisa. 2016. *Placing outer space: An earthly ethnography of other worlds*. Durham: Duke University Press.
Packer, Joseph. 2015. *Alien life and human purpose: A rhetorical examination through history*. Lanham: Lexington Books.
Reeves, Eileen. 2014. Astrology and literature. In *A companion to astrology in the Renaissance*, ed. Brendan Maurice Dooley, 287–331. Leiden: Brill.
Roberts, Adam. 2006. *The history of science fiction*. Basingstoke: Palgrave Macmillan.
Rogers, Brett M., and Benjamin Eldon Stevens, eds. 2015. *Classical traditions in science fiction*. Oxford/New York: Oxford University Press.
Roos, Anna Marie Eleanor. 2001. *Luminaries in the natural world: The sun and the moon in England, 1400–1720*. New York/Oxford: Peter Lang.
Suvin, Darko. 2016 [1979]. *Metamorphoses of science fiction: On the poetics and history of a literary genre*. Oxford: Peter Lang.
Vermij, Rienk, and Hiro Hirai. 2017. *The marginalization of astrology: Introduction. Early Science and Medicine* 22: 405–409.
Whewell, William. 1853. *Of the plurality of worlds: An essay*. London: J.W. Parker and Son.
Wolloch, Nathaniel. 2000. Christiaan Huygens's attitude toward animals. *Journal of the History of Ideas* 61: 415–432.
———. 2002. Animals, extraterrestrial life and anthropocentrism in the seventeenth century. *The Seventeenth Century* 17: 235–253.
Wood, Chauncy. 1970. *Chaucer and the country of the stars: Poetic uses of astrological imagery*. Princeton: Princeton University Press.
Young, Edward. 1813. *Works*, 3 vols. London: F. C. and J. Rivington.

Index

© Springer Nature Switzerland AG 2019

J. E. Christie, *From Influence to Inhabitation*, International Archives of the History of Ideas Archives internationales d'histoire des idées 228,
https://doi.org/10.1007/978-3-030-22169-0

9783030221713